广州白云国际机场
二号航站楼及配套设施

GUANGZHOU BAIYUN INTERNATIONAL AIRPORT
TERMINAL 2 AND SUPPORTING FACILITIES

主编：陈雄、潘勇、周昶
CHIEF EDITORS: CHEN XIONG, PAN YONG, ZHOU CHANG

中国建筑工业出版社

序言

何镜堂

中国工程院院士
全国工程勘察设计大师

2018年4月26日,广州白云国际机场二号航站楼及配套设施工程正式启用。二号航站楼的启用提升了广州白云机场的国际枢纽竞争力,强化了广州地区的综合交通枢纽功能,完善了我国民航运输网络,是实现建设世界级机场群目标的重要举措,提升了广州城市国际影响力与竞争力,进一步巩固广州作为国家中心城市和粤港澳大湾区综合性门户城市的地位,为建设具有国际影响力的现代化大都市迈出了重要一步。

作为广州迈向未来的城市新门户,建成后的二号航站楼在形象塑造、规划构型、空侧规划、陆侧交通、流程工艺、空间设计、环境营造等方面都达到了较高的水平,反映了中国最新的建筑技术水平,展示了新时代中国大型标志性建筑的先进性与独特性。

二号航站楼是一个优秀的建筑作品,其建筑形象个性鲜明,设计风格灵动清新。规划构型简洁灵活,流程布局清晰合理,使用管理便捷高效。空间设计依托流程,风采各异恢弘灵动,带给旅客强烈的有序变换的视觉冲击力,塑造了具有岭南地域特色的门户形象。建筑结合气候及环境,结合庭院与自然,体现了岭南建筑地域风格特征,体现了整体观和可持续发展观,地域性、时代性和文化性的岭南建筑创作理念。室内装修风格简洁明快,内部环境清雅舒适,彰显了大型交通建筑的神韵。同时在航站楼的设计过程中结合节能环保理念打造了绿色生态机场。

二号航站楼启用一年多以来,为南来北往的客人提供了优质的服务,其造型、空间、环境、设施给人们留下了难以忘怀的印象,获得建筑行业、民航行业及社会各界的广泛好评。二号楼的建成标志着广州正面向国际、面向未来,以坚定的信念、无比的干劲,实现现代化大都市梦想,助力粤港澳大湾区发展与建设中国成为世界民航强国的伟大战略目标!

本书的出版,将是对中国民航、对广东省广州市这一重大工程诞生的很好的纪念和总结,这本技术全面且图文并茂的专著也是我国为超大型航站楼设计而出版的少有的大型著作。

应全国工程勘察设计大师、广东省建筑设计研究院陈雄副院长兼总建筑师之邀,欣然命笔,谨以此文作序。

FOREWORD

He Jingtang

Academician of Chinese Academy
of Engineering (CAE)
National Engineering Survey and
Design Master

On April 26, 2018, Terminal 2 (T2) and Supporting Facilities of Guangzhou Baiyun International Airport (Baiyun Airport) was officially put into operation. The operation of T2 enhanced the competitiveness of the Baiyun Airport as an international hub and Guangzhou's function as a comprehensive transportation hub, and improved China's civil aviation transportation network. As an important move to create a world-class airport cluster, the Project boosted the city's international presence and competitiveness and strengthened its position as a central city in China and an integrated gateway city in Guangdong-Hong Kong-Macao Greater Bay Area (GBA), signifying a major step towards the goal of making Guangzhou a modern metropolis with international influence.

As the city's new gateway to the future, the T2 complex, upon its completion, has reached an impressive level in appearance, planning layout, airside planning, landside traffic, flow design, spatial design and environmental creation. It represents the latest architecture and construction technology in China and serves as an advanced and unique landmark of the country in the new era.

T2 is a marvelous work of architecture. It features distinctive architectural image, refreshing design style, concise and flexible planning layout, clear and reasonable flow design, and convenient and efficient use and management. Its flow-based spatial design creates spectacular spaces of varied features, bringing strong visual impacts to passengers while shaping a gateway image with distinctive Lingnan characteristics. Well responding to the local climate and environment, T2 integrates with the courtyards and natural elements to present the local characteristics of Lingnan architecture. It is a perfect example that embodies the concepts of holism and sustainability, as well as the features of place, culture and time. The interior finish of T2 is concise, elegant and comfortable, conveying its inner beauty as a gigantic transportation building. Energy conservation and environmental protection concepts are incorporated in the design to create a green and ecological airport.

T2 has offered high quality service to passengers from around the world over the past year since its opening. It has earned a reputation in the building and aviation industries and among the public for its impressive form, space, environment and facilities. Its completion signifies that Guangzhou, with a future- and world-oriented mindset, is working confidently and vigorously towards the goal of making itself a modern metropolis, driving the growth and development of GBA, and turning China into a leader in the global civil aviation industry.

The publishing of this book would be a perfect commemoration and recap of this centurial project for China's civil aviation industry, for Guangdong Province and for the city of Guangzhou. With abundant images and detailed descriptions, it would be one of the very few comprehensive publications in China elaborating on the design of a mega terminal project.

It is a great honor for me to write this foreword for the book at the invitation of Mr. Chen Xiong, a National Engineering Survey and Design Master, and the vice president and chief architect of the Architectural Design and Research Institute of Guangdong Province (GDAD).

目录

012		序言
016	**第1章**	**总论**
032	**第2章**	**图纸**
034	2.1	效果图
044	2.2	总平面图
046	2.3	各层平面图
052	2.4	立面图、剖面图
062	**第3章**	**工程概述**
064	3.1	机场定位
064	3.2	航站区现状及发展预测
068	3.3	机场总体规划与本期扩建工程
072	3.4	工程设计亮点
082	**第4章**	**构型设计与交通组织**
084	4.1	方案推演
086	4.2	空侧规划保证飞机运行效率
087	4.3	与空侧容量相匹配的陆侧交通设计
090	**第5章**	**流程设计与功能配置**
092	5.1	概述
093	5.2	功能分布与流程设计
100	5.3	功能配置

102	**第6章**	**建筑设计**
108	6.1	设计构思
109	6.2	流动的造型
110	6.3	外围护结构系统设计
126	6.4	主要空间衔接与室内设计
152	6.5	大空间照明设计
156	6.6	色彩体系设计
160	6.7	标识系统设计
168	6.8	广告设计
170	6.9	文化设计
172	6.10	主要旅客功能设施设计
196	6.11	景观设计
198	6.12	绿色建筑设计
204	**第7章**	**技术设计**
206	7.1	行李系统设计
212	7.2	结构设计
220	7.3	机电设计
236	7.4	市政工程设计
246		**工程大事记**
248		**广东省建筑设计研究院项目设计团队**
255		**项目科研及成果**
260		**项目有关数据**
261		**本书编辑团队**
262		**编后记**

CONTENTS

013	FOREWORD	100	5.3 FUNCTIONAL CONFIGURATION
016	**CHAPTER 1 OVERVIEW**		
		102	**CHAPTER 6 ARCHITECTURE**
032	**CHAPTER 2 DRAWINGS**		
		108	6.1 CONCEPTION
034	2.1 RENDERING	109	6.2 FLOWING FORM
044	2.2 MASTER PLAN	110	6.3 BUILDING ENVELOP
046	2.3 FLOOR PLANS	126	6.4 MAIN SPACE CONNECTIONS AND INTERIOR DESIGN
052	2.4 ELEVATIONS AND SECTIONS	152	6.5 LARGE-SPACE LIGHTING
062	**CHAPTER 3 PROJECT PROFILE**	158	6.6 COLOR SYSTEM
		160	6.7 SIGNAGE SYSTEM
065	3.1 AIRPORT POSITIONING	168	6.8 ADVERTISEMENT
065	3.2 EXISTING CONDITIONS AND DEVELOPMENT PROJECTION	170	6.9 CULTURAL FEATURES
070	3.3 MASTER PLAN AND EXPANSION PROJECT	172	6.10 FUNCTIONAL FACILITIES FOR PASSENGERS
072	3.4 DESIGN HIGHLIGHTS	196	6.11 LANDSCAPE ARCHITECTURE
		198	6.12 GREEN BUILDING
082	**CHAPTER 4 TERMINAL LAYOUT AND TRAFFIC ORGANIZATION**	204	**CHAPTER 7 TECHNICAL SYSTEMS**
084	4.1 EVOLUTION OF SCHEMES	208	7.1 BAGGAGE
086	4.2 AIRSIDE PLANNING FOR EFFICIENT AIRCRAFT OPERATION	213	7.2 STRUCTURE
088	4.3 LANDSIDE TRAFFIC COMPATIBLE WITH AIRSIDE CAPACITY	222	7.3 MEP
		238	7.4 UTILITIES
090	**CHAPTER 5 TERMINAL FLOW AND FUNCTIONAL CONFIGURATION**	247	**MILESTONES**
		249	**DESIGN TEAM OF GDAD**
092	5.1 GENERAL	256	**RESEARCHES & AWARDS**
093	5.2 FUNCTIONAL DISTRIBUTION AND FLOW DESIGN	260	**FACTS & DATA**
		261	**EDITORIAL TEAM**
		263	**AFTERWORD**

第1章

总论

CHAPTER 1 **OVERVIEW**

总论

创新、先进、独特的
新岭南门户机场航站楼

—— 广州白云国际机场二号航站楼及配套设施工程创作实践

陈雄

全国工程勘察设计大师
广东省建筑设计研究院副院长、总建筑师

广州白云国际机场是全国三大国际枢纽机场之一，是粤港澳大湾区航空枢纽最重要的组成部分，是广东省"5+4"机场群的核心。白云机场扩建工程（二号航站楼及配套设施）是"十二五"规划和《珠三角规划纲要》重点建设项目。

二号航站楼是超大型国际枢纽航站楼，设计年旅客量4500万人次，本期扩建总机位78个，其中近机位64个，二号航站楼建筑面积65.87万平方米，为四层混凝土大跨度钢结构建筑。配套交通中心及停车楼建筑面积20.84万平方米，为地下二层地上三层的钢筋混凝土结构建筑。

二号航站楼以旅客体验为导向，充分满足旅客需求的国际航空枢纽为目标进行设计，在规划构型、流程布局、空间造型、绿色节能、商业设施、地域特色、技术设计、智慧机场等多方面进行了创新设计，追求先进、特色鲜明。

1. 赋予建筑灵魂的"云"概念

从一号航站楼开始，建筑师并没有刻意提出一个具体的符号或主题，只是强调一种现代、简洁、流畅、精致的设计风格。随着二号航站楼的创作，建筑师抽取白云元素，形成"云"概念。"云"概念构思包括多重内涵：

（1）"云"概念，源自白云，与"白云国际机场"天然契合，也预示了旅客即将开始飞行于蓝天白云的梦幻旅程。
（2）在当今世界的语境中，"云"概念就是云端的连接，借此也表达了二号航站楼是旅客连接世界的地方。
（3）"云"概念借鉴了当今最新科技云计算的理念，喻示着二号航站楼的设计与建设将集成当今时代的创新科技和理念，致力于为旅客带来良好出行体验。
（4）"云"概念与机场航站楼的功能特征非常吻合，飞行就是旅客群体的流动，交通建筑的效率与便捷非常重要。

2. 实现"平安、绿色、智慧、人文四型机场"的设计策略

（1）创建布局合理、功能齐备、流程顺畅、便捷高效的平安航站楼。合理规划流程及功能布局，确保旅客出行与运行管理效率。努力建构服务设施及安全保障设施体系，确保服务标准与安保水平。
（2）创建地域特色显著、环保生态的绿色航站楼。充分发掘地域特征及传统文化特色，综合采用多种绿色建筑技术，力求达到资源、能源的最大化利用，创造高效、健康、节能、舒适的新型绿色机场建筑，促进人与自然、环境与发展、建设与保护相平衡的航站楼体系。
（3）创建科技领先、智能高效的智慧航站楼：采用航班动态通知、自助行李托运、DCV小车自动行李分拣、生物技术识别等先进技术，打造全流程的智慧体系，以智能科技引领高效服务。
（4）创建主题突出、内涵丰富的人文航站楼。将底蕴深厚的华夏文明与特色鲜明的岭南文化融入设计，通过文化设施建构、文化展示 区域预留、文化特征融入设计内容等方式，充分展示航站楼的人文气息。
（5）创建外形磅礴、空间流畅、景致宜人的现代化门户航站楼。承接并拓展一号航站楼的标志性设计元素，塑造现代大气的外形的特征，保持主要空间的特色与衔接，营造步移景异的空间景观效果，体现标志性的门户形象。
（6）创建设置灵活、弹性预留的可持续性航站楼。结合布局、流程及机场未来发展需求，预留增设功能及发展建设空间。

3. 满足枢纽型机场核心工艺的航站楼构型

二号航站楼采用"指廊式＋前列式"混合构型，具有飞机停靠面长、近机位多、近距离大机位多、旅客步行距离适中、国际国内互换方便等优点。

3.1 空侧规划保证飞机运行效率

（1）空侧规划原则。适应航站楼的设计整体，充分利用现有土地资源；采用南航所提供未来机队的机型组合，满足其发展需要；提供最高效的机坪与机位配置，以满足南航作为基地与枢纽航空公司的运营需求。
（2）混合机位概念。航站楼国内与国际运营高峰小时不同，为提高机位的使用效率，设计引入可在国内与国际运营之间进行转换的混合机位概念，在流线最便捷、流程最短的前列式机位设置成国内/国际可切换的混合机位，提高机位使用灵活性及旅客处理效率，满足国际业务快速增长的需要。
（3）组合机位概念。为实现机位布置最大化与提升空侧机坪使用灵活度，设计引入可供大飞机与小飞机同时停靠的组合机位，即可供一架E类、F类飞机或两架C类停靠的机位，达到机位满足各类飞机错时靠港的使用需求。
（4）保证近机位数量。二号航站楼一期工程的飞机停靠

总论

面达3545m,包括64个近机位,其中国内机位共34个(31C、3E),国际机位共21个(8C、11E、2F),国内与国际可转换的混合机位共9个(7E、2F);二期建设中扩建的东四指廊和西四指廊将再增加26个近机位,总体规划共90个近机位。

(5)双滑行道的指廊间距。指廊与指廊之间按照同类机型布置机位,用双滑行道保证滑行效率,并节省空间。

3.2 与空侧容量相匹配的陆侧交通设计 陆侧交通一体化,实现城轨、地铁、大巴、出租车等各种交通方式无缝连接,平层解决主要交通换乘,缩短换乘步行距离

(1)南北贯通道路交通。航站区陆侧设置了贯通南北的道路系统,确保白云机场与外部城市交通衔接的安全性。通过东西环路实现一号、二号航站楼分别与南北出入口的快捷联系,两楼的交通互不干扰,减少车辆长距离绕行给机场环路带来的交通压力,达到了既有效分流车辆,又保障道路畅通的目的。

(2)交通中心及停车楼。交通中心为旅客换乘各种交通工具提供了方便。交通中心以"公共交通优先、大众旅客优先、离港旅客优先"为设计原则,公共交通(地铁、城轨、市区大巴、长途大巴、出租车)前置,缩短了大部分旅客步行距离。地铁站厅和城轨站厅分列交通中心大厅的东西两侧,有效分流旅客。私家车停车楼紧邻交通中心,旅客可通过垂直交通到达各层车库。

(3)与轨道交通衔接。白云机场配套了两条地下轨道即地铁三号线和城际轨道线,均在交通中心地下二层设置站厅。地铁、城轨的出发旅客和到达旅客与航站楼均有各自的流线设计,互不干扰;同时,两轨道交通之间设有专用的换乘通道,高效便捷。

(4)人车分流的陆侧交通。二号航站楼到达层(首层)迎客厅与交通中心之间设置开阔的行人广场,而航站楼前的到达大巴车道及出租车道均采用下穿隧道的方式,实现完全的人车分流,旅客可以方便地平层步行到达市区大巴上客区、长途大巴候车区、私家车停车楼,或通过垂直交通到达地铁站厅、城轨站厅及停车楼其他楼层。

4. 旅客流程与行李流程为主线的航站楼工艺

旅客流程设计和行李系统工艺设计是实现航站楼交通建筑功能的基础核心设计,它决定了航站楼内部功能组织框架,是影响航站楼运行效率的最重要因素之一。

4.1 旅客流程设计

作为一个大型的复合枢纽机场航站楼,二号航站楼拥有各种类型的旅客流程,包括国内出港、国内进港、国际出港、国际进港、国内转国际、国内转国内、国际转国内、国际转国际、国际航班国内段(国内-广州-国际、国际-广州-国内、广州-国内-国际、国际-国内-广州)等,各流程或交叉或联系或结合又形成新流程,错综复杂。我们采取了国内混流、国际分流、设置混合机位的策略来搭建航站楼的内部组织框架和楼层。

(1)主流程旅客步行距离短。从车道边下车后,到最远登机口步行≤700m,步行时间<9分钟。

(2)单一方向的旅客流程。在流线设计上采取了单一方向、减少选择点的策略。旅客进入办票大厅后只有向前一个方向,不需做选择即到达安检厅,安检后开始选择登机口方向。到达旅客也是从各个指廊汇聚到主楼提取行李,国际和国内各有一个迎客口,统一的到达厅与交通中心连通,空间结构简单、方向引导非常清晰。

(3)国内流程采用混流设计。到达旅客下登机桥与出发旅客混合后去往行李提取厅。国内混流设计较出发、到达分流设计既减少了独立的到达楼层通道和服务机房面积,又共用了商业、卫生间、问询柜台等服务设施,还减少了登机口管理人员,降低了投资,提高了资源利用率。

(4)国际流程仍采用分流设计。因海关和边防检查监管要求,采用了传统的出发、到达旅客分流的模式。

(5)国内/国际可切换的混合机位流程。国际和国内旅客可同时分层候机,通过登机桥固定端实现功能衔接与切换,提高了机位使用率。

(6)便捷的中转流程设计。中转流程是二号航站楼最重要的流程之一,其设计非常富有特点。国内转国内、国际转国内联程旅客到达、中转和候机全程在二层平层解决,国内转国际、国际转国际旅客流程经过主楼中转通道汇聚后,通过竖向交通引导至三层国际始发旅客联检口。中转流程检查共用始发、终到流程的检查场地资源,节约场地和人力资源,降低标识引导的复杂性,减少旅客选择点。

(7)国际国内并置的安检厅。在机场的检查流程上,二号航站楼最大的特点是将国际、国内安检并排设置在第一关,利用国际、国内高峰小时错峰的特点,对安检通道国际国内功能进行切换,解决特殊高峰时段(如春节)流量瓶颈问题。

(8)规划APM系统服务长距与中转流程。二号航站楼采用"前列式+指廊式"的构型,部分指廊离主楼较远。为解

决步行距离过长、一号、二号航站楼旅客中转和楼层转换，二号航站楼内设置预留了两个APM站。一个站为国内集中出发、到达站，另一个为国际-国内、国内-国际中转站。旅客可通过APM系统快捷去往较远的指廊和一号航站楼候机，同时也可利用APM系统快捷进入行李提取流程和中转检查流程。

4.2 行李系统及技术协调

航站楼行李系统由接收行李处理子系统、行李安全检测子系统、行李自动分拣捷运子系统、行李自动暂存子系统、特殊行李处理子系统、提取行李传送子系统等多项子系统构成。在核心的自动分拣捷运子系统工艺技术上，采用国内首个DCV自动行李分拣系统，分拣速度快、出错率低、效率高。行李机房设置在首层与机坪平层行李车在机坪与行李机房间可快捷的平层高效运行。

5. "云"概念下的原创性造型与空间

从白云轻盈、漂浮、流动的特质出发，定义了二号航站楼设计的整体氛围和协调性，塑造出流动的造型特色、轻盈的空间表现与柔和的空间格调，体现在以下几个方面：

（1）从"云"概念，提炼出连续云状拱形的造型母题。连续多跨拱形的张拉膜雨篷，与航站楼南立面檐口连续拱形曲面在造型上下呼应，膜材漫反射的柔光映衬檐口的金属质感，形成层次丰富、光影变幻、连绵起伏的"云"意象。
（2）源自于"云"概念的简洁且流畅的曲面造型整合了主楼与东西两翼的不同建筑体量，使之融合在一起，最终实现与周边自然环境的契合。
（3）"云"概念元素体系化层进式表达，由建筑外观造型到室内空间，由外而内，犹如行云流水，层层递进，一气呵成。从出发车道边到办票大厅，由三条通廊一直贯穿到安检大厅，最终延伸到候机指廊，从室外城市空间到室内公共空间，"云"概念元素在建筑造型与空间中流动转换，节奏灵动流畅，塑造出具有原创性的二号航站楼建筑与空间形象。
（4）契形线条及体量对流动性的表达，设计上引入交汇的契形线及体量，有别于正交形体，由于不平行的契形线在透视上的错觉，在空间上具有不稳定感，从而带来很好的流动感。
（5）引入自然光线，光影赋予空间流动性，同时自然地引领旅客，让旅客身心舒畅。
（6）以共享空间的手法营造重要空间节点，衔接交通空间与商业空间，也充满流动性。
（7）保留了弧线形的主楼、Y形柱及张拉膜雨篷这些白云机场特有的元素，与一号航站楼建筑造型和谐一致，共同建构了"双子星"航站楼的完整形象。

6. 实现建筑造型的屋面系统和幕墙系统设计

6.1 金属屋面系统

金属屋面总面积约26万平方米。拱形非线性屋面造型连续流畅，主楼和连接楼相连一气呵成。全部采用外天沟以避免漏水。采用铝镁锰合金直立锁边屋面板系统，主要构造包括铝镁锰合金屋面防水板、玻璃纤维保温层、柔性防水层、纤维水泥板支撑层、岩棉支撑层、隔汽层、压型钢底板支撑层、主次檩条。这种系统是成熟的技术，满足造型找形、隔热保温、排水防水、自然采光各种功能需求，施工周期短，施工和维护都很方便，造价可控。为加强金属屋面的抗风性能，采用三道措施进行加强。主楼屋面布置一排整齐的条形天窗，指廊屋面的中间布置一条天窗。天窗造型简洁实用并兼顾屋面造型的美观。采用钢化夹胶中空玻璃（Low-E）。

6.2 幕墙系统设计

玻璃幕墙总面积约10万平方米，其中最大尺度幕墙为主楼正立面，总面宽约430m，最大高度36m。与点支式玻璃幕墙的一号航站楼不同，二号航站楼采用横明竖隐的玻璃幕墙系统，玻璃板块尺寸为3m×2.25m，高度分格与建筑层高模数2.25m相同，宽度分格与建筑平面模数相同。二次结构采用立体钢桁架为主受力构件，横向铝合金横梁为抗风构件，竖向吊杆为竖向承重构件。铝合金横梁与水平遮阳板一体采用挤压铝型材。竖向吊杆藏于玻璃接缝中，整个幕墙单元具有通透感。立体钢桁架结构无需水平侧向稳定杆，幕墙整体感强、通透美观。设备管线与幕墙二次结构相结合，将管线埋藏于U型结构内，正面再用铝扣板遮挡。为配合消防排烟开启扇角度达到60°的需要，采用可以大角度开启的气动排烟窗。考虑夏热冬暖地区的节能需要，西侧幕墙外侧设置机翼型电动可调遮阳百叶。有0°、30°、60°和90°四个角度。百叶选用微孔板，可以达到遮阳且透影的效果。

7. 建筑、装修与结构的一体化设计

7.1 主楼采用钢管柱+网架体系

与一号航站楼相比，结构体系做了重大调整，采用普通成熟的网架体系，办票大厅柱网横向为36m，纵向为53m+45m+54m，在纵向大跨处做加强网架处理，更加经济且满足工期要求。采用钢管柱以缩小截面，并且把大厅中间的两排柱子收藏在12个办票岛里，在旅客的主要活动范围里完全没有柱子阻挡。指廊柱网纵向18m，横向35m和45m，采用预应力混凝土柱，缩小截面，优化空间效果。

7.2 全国首创旋转渐变的"云"概念天花

装修设计与建筑的动感造型一气呵成，同样表达出"轻盈、漂浮、流动"的感觉，而且与结构形状相适应，使旅客在室内能够体验开敞明亮舒适的半室外空间感受。以办票大厅的吊顶为例，采用全国首创旋转渐变的天花。36m为单元的波浪吊顶造型，简洁而富有动感，包络了结构的加强网架，以3000mm长、500mm宽、50mm高的条形铝合金板每5度渐变旋转定位安装，在条形天窗下尽可能通透，没有天窗的部位则是平顶，既解决了天窗采光要求，也避免了完全暴露网架，形成完整而富有动感的室内空间界面。

7.3 具有航空器动感外形的办票岛设计

结合航空飞行器的特点，营造动感且具冲击力的非线性岛体外形办票岛。综合协调建筑、结构、机电、弱电、行李与安检安保各专业满足办票岛的复合功能。

7.4 功能流程齐全与造型独特的登机桥

设置国际国内"可转换机位"与"登机桥固定端"（含A380），满足南航未来机队的机型组合系统及发展需要，引入可在国内与国际运营之间进行转换的混合机位概念，将流线最便捷、流程最短的前列式黄金大机位设置成国内/国际可切换的混合机位，通过登机桥固定端实现机位与航站楼各进出港层的对接，提高机位使用灵活性及旅客处理效率，满足航司业务快速增长的需要。

外形设计延续了航站楼的设计手法，大胆对形体进行了斜切，结合立面铝板斜线分隔，延续了航站楼的动感造型特色，使之与航站楼主体有很好的体量过渡，也有效减少了空间体积而有利节能。

总论

跨度为18-24m，钢结构采用巨型刚桁架体系，设计要求竖腹杆与上下弦杆刚接，两榀巨型刚桁架之间采用钢框架梁连接。

7.5 特有的张拉膜雨篷设计

张拉膜雨篷是白云机场特有的设施和标志之一，适合岭南地区的多雨炎热气候。从一号航站楼到二号航站楼的车道边都覆盖了张拉膜雨篷，为旅客遮风挡雨遮阳，可以在不好的天气下从容上下车。

膜结构设计采用骨架膜结构，骨架采用钢结构，重点部位采用铸钢节点，膜材采用PTFE膜材。

8. 表达特色鲜明的地域文化

在三层、五层的岭南花园，让旅客在现代化的航站楼内可以感受到传统园林的魅力，是人与自然融合的最佳实践，身处这里，犹如漫长旅途中的一片绿洲，体现了广州地处中国南方的气候特点。岭南花园属于狭长式用地，设计采用斜线条式的铺装布局，使空间变得生动有趣，配以造型树池、水景、汀步等穿插布局，以达到小中见大，空间层次分明的效果。北指廊的国际到达旅客可以欣赏到岭南花园的美景，对广州留下美好的第一印象。

"宇宙飞船"——办票大厅的文化广场，彰显独特的公共文化魅力，成为展示公共文化艺术和商业的独特平台。也可以举办商业推广活动。岭南花园、文化广场、时尚及传统多维的体验、公共艺术、独特的标志性和文化内涵、流畅的旅客体验、时空隧道使旅客体验时空穿梭的绚丽感觉，共同构建机场文化特色。

9. 融合旅客流程的非典型商业综合体

非典型商业综合体概念：根据对国际国内机场不同案例及对商业中心的研究，提出了独特的航站楼商业概念——非典型商业综合体。航站楼的重中之重是旅客流程，旅客聚散集中，旅客流动性高，停留时间短。因此，为了达到增加非航收入的目标，在和商业顾问共同研究之后，从多方面创造了不同于国内其他枢纽机场的航站楼商业布局模式：整合旅客流程，合并商业动线；创造跃层中庭，营造商业氛围；组合功能布局，控制业态配比；引入生态绿化，优化商业环境。结合流程巧妙融入商业设施，使旅客在出行途中可享便利的购物环境，增加非航收入。

9.1 整合旅客流程，活跃商业动线

将可以合并的流线整合一体化，减少消极空间流线。调整垂直交通，减少重复的交通面积。将商业流线与旅客流线紧密结合，增加商业与旅客的接触面，杜绝阴阳铺。

9.2 协调专业资源，合理增加面积

综合协调建筑、结构、机电等专业。量体裁衣，节省结构机电等富余空间。创造更多商业面积，增加航站楼商业价值。

9.3 创造跃层中庭，营造商业氛围

整合平面功能，在航站楼中创造传统Mall的垂直商业中庭，提高空间汇聚性，增加商业氛围，也可增加各类广告的可视面。

9.4 组合功能布局，控制业态配比

根据预测数据合理设置商业餐饮与零售配比，保证未来餐饮、零售功能的可转换性。优化业态布局，平衡功能需求与商业需求，增加商业设施的有效性。

9.5 引入生态绿化，优化商业环境

在旅客重要流程及商业活动区设置绿化中庭或花园，优化航站楼室内环境质量，进一步提高商业价值，体现岭南建筑的特色。

10. 节能环保为导向的绿建设计

航站楼空间体量巨大，外围护结构材料选择与设备选型是建筑节能的关键。二号楼采用各种绿色节能设计措施，取得国家"三星级绿色建筑设计标识证书"，成为湿热地区首个三星级绿色大型公共交通枢纽建筑，使项目成为航站楼绿色建筑的典范。

10.1 理性的热环境分析

采用CFD流体力学软件和光环境分析软件对建筑平面设计和玻璃幕墙 可开启位置及面积进行确定，通过节能计算对屋面材料与构造进行优化。采用动态能耗模拟软件DeST和Daysim照明能耗模拟软件对建筑全年空调采暖能耗和照明能耗进行分析。

10.2 合理的自然采光

结合空间尺度及平面功能对项目中的典型功能空间进行自然采光专项设计，航站楼办票大厅上空屋面均匀设置采光天窗。每年节约用电量约280万度，约占照明总用电量20%。交通中心地下停车场设置采光井改善地下空间的自然采光。

10.3 组织自然通风

广州市年平均风速在2m/s左右，风力资源不够丰富。采用CFD流体力学软件对建筑平面设计和玻璃幕墙可开启位置及面积进行指导优化，加强建筑的自然通风能力，减少建筑夏季空调能耗。对于位于航站楼内部联检区等通风条件较差的区域，通过设置可供旅客休憩的中庭花园以加强区域的自然通风条件，改善建筑用能效率和提高室内空气品质。

10.4 高效的太阳能利用

广州地区太阳高度角较大，太阳辐射总量与日照时数均充足。充足的太阳辐射和日照为太阳能的利用创造了良好的条件。二号航站楼的计时旅馆、头等舱及商务舱区域采用太阳能加热泵热水系统，太阳能热水用量占建筑生活热水消耗量的72%；航站楼安检大厅上方金属屋面采用太阳能光伏发电。

10.5 节能舒适的遮阳设计

二号航站楼西侧选用机翼型电动可调遮阳百叶，东侧采用室内电动遮阳卷帘。

结语

大型航站楼建筑设计是多学科融合与创新的典型例子，需要建筑师深入理解航站楼各方使用的需求，还要与其他各专业顾问密切配合，必须融合众多学科及复杂的技术体系，并在此基础上进行大力创新，通过多学科多专业协同一体化设计，才能够实现航空门户交通枢纽的高水平目标。二号航站楼自通航以来，获得业界及社会的广泛好评，荣获SKYTRAX "全球五星航站楼"、全球最杰出进步机场及"中国最佳机场员工"等奖项。广州白云国际机场是南方航空总部基地，承担起祖国南大门连接欧、美、澳、非的空中桥梁。崭新的二号航站楼与一号航站楼交相辉映，助力广州建设粤港澳大湾区核心城市，以及中国建设成为世界民航强国的伟大战略目标！

OVERVIEW

AN INNOVATIVE, ADVANCED AND UNIQUE AIRPORT TERMINAL AS A NEW GATEWAY TO LINGNAN REGION

Design Practice for Terminal 2 and Supporting Facilities of Guangzhou Baiyun International Airport

Chen Xiong

National Engineering Survey and Design Master
Vice President of GDAD, Chief Architect

As one of the three major international hub airports in China, Guangzhou Baiyun International Airport (Baiyun Airport) is the most important part of the aviation hub in Guangdong-Hong Kong-Macao Greater Bay Area (GBA), and the core of "5+4" Airport Cluster in Guangdong Province. Baiyun Airport Expansion Project (Terminal 2 and Supporting Facilities) is a key development project in the 12th Five-Year Plan and *Outline Development Plan for the Pearl River Delta*.

Terminal 2 (T2) is a mega international hub terminal with a designed capacity to handle 45 million passengers annually. The expansion project intends to add a total of 78 stands, including 64 contact stands. T2 is a four-floor concrete building with large-span steel structure and a GFA of 658,700 m². The Ground Transportation Center (GTC) and parking building, with a GFA of 208,400m², is a reinforced concrete structure with 2 basement floors and 3 above-grade floors.

Envisioned as an international aviation hub oriented towards passengers' experience and full needs, T2 comes with innovative design in planning layout, flow, spatial design, green building, energy conservation, commercial facilities, regional features, technical design, smart airport etc., pursuing advanced design with unique features.

1. Cloud Concept as the Soul of the Building

The design of Terminal 1 (T1) emphasizes a modern, concise, smooth and exquisite design style instead of a specific symbol or theme. For T2, the concept of cloud is proposed based on the element of white cloud. The concept of cloud incorporates many meanings:

(1) Inspired by white cloud, it naturally matches the name of Baiyun (i.e., white cloud) International Airport, and signifies the fantastic journey above the white clouds in the blue sky.
(2) In a context of the modern world, it means cloud connection, highlighting T2 as passengers' connection to the world.
(3) It also means the modern technology of cloud computing, signifying the integration of modern innovative technologies and ideas in the design and development of T2 to bring the best travel experience for passengers.
(4) It perfectly coincides with the functions of an airport terminal, as air flights represent the collective passenger flows so the importance of efficiency and convenience of a transportation building should never be neglected.

2. Design Strategy to Create a Safe, Green, Smart and Cultural Airport

(1) Establish a safe terminal with rational layout, complete functions, smooth flows, convenience and high efficiency. Rationally organize flows and functional layout for efficient passenger travel as well as operation and management. Establish a service facility and safety assurance system to ensure service standards and security level.
(2) Establish a green terminal with distinct regional features that is friendly to the environment and ecology. Fully explore regional and traditional cultural features, integrate green building technologies to maximize the utilization efficiency of resources and energies, create a highly-efficient, healthy, energy-saving, and comfortable new green airport building, and facilitate a coordinated terminal system featuring harmony of man and nature, balance between environment and development, construction and protection.
(3) Establish a smart terminal featuring leading

OVERVIEW

technologies, intelligence and high efficiency. Provide real-time flight notification, self-service baggage check-in, DCV (Destination Coded Vehicle) baggage handling system, biotechnological identification etc., to realize a whole-process smart system based on science and technology for highly efficient services.
(4) Establish a cultural terminal highlighting the theme and profound cultural connotation. Integrate the profound Chinese civilization and featured Lingnan culture into design, and fully present the cultural aspect of the terminal by providing cultural facilities, reserving cultural exhibition area and integrating cultural features into design.
(5) Establish a modern gateway terminal with spectacular appearance, flowing spaces and attractive landscape. Continue and expand the iconic design elements of T1 and create a modern and generous appearance, maintain the features of major spaces and interconnections to realize varied landscapes along the way and showcase the landmark identity of the gateway.
(6) Establish a sustainable terminal with flexible and reserved functions. Based on layout, flows and future development needs, reserve additional functions and space for future development.

3. Terminal Layout Meeting Core Technical Needs of a Hub Airport

T2 adopts a mixed "pier + linear" layout which enjoys the advantages of longer parking frontage, more contact stands, more contact stands for large aircrafts, proper walking distance for passengers, convenient transfer between domestic and international flights etc.

3.1 Airside planning to ensure aircraft operation efficiency

(1) Airside planning principles.Adapt to the holistic terminal design to fully utilize existing land resources, and reference the future aircraft mix provided by the China Southern Airlines (CSN) to meet its development needs; provide the most efficient apron and stand configuration to meet CSN's operation needs as a base and hub airline.
(2) The concept of swing stands. In view of different domestic and international peak-hours, to improve stand efficiency, the concept of swing stands is introduced to enable conversion between domestic and international operations. Linear stands with the most convenient and shortest flows are designed into swing stands to enhance flexible stand operation and efficient passenger handling and meet the demands of rapidly growing international airline business.
(3) The concept of MARS (Multiple Aircraft Ramp System) stands. To maximize the number of stands and the flexibility of airside apron operation, MARS stands are introduced in design for the parking of both large and small aircrafts. Each MARS stand can accommodate a Code E aircraft and a Code F aircraft or two Code C aircrafts at the same time, allowing various codes of aircrafts to park during staggered time periods.
(4) Sufficient contact stands.For T2, the aircraft parking frontage in Phase I measures 3,545m to accommodate 64 contact stands, including 34 domestic stands (31C, 3E), 21 international stands (8C, 11E, 2F) and 9 swing stands (7E, 2F). In Phase II expansion project, another 26 contact stands are added to the East 4th Pier and the West 4th Pier, making a total of 90 contact stands.
(5) Pier interval for double taxiways.Pprovide stands for the same aircraft type between piers, with double taxiways to ensure the efficiency of taxiing and space.

3.2 Landside traffic design matching airside capacity: provide integrated landside

transportation to realize seamless interchange between all means of transportation including intercity railway, metro, bus, taxi etc., with major interchanges organized on the same level to reduce walking distance

(1) North-south road traffic.A north-south road system is provided on landside of the terminal area to provide safe traffic connection between the Baiyun Airport and external urban transportation. Provide convenient north and south access for T1 and T2 respectively via east and west ring roads to avoid mutual interference and reduce the pressure of long detour on airport ring roads. This way the vehicular traffic can be effectively diverted and smooth road conditions can be guaranteed as well.
(2) GTC and parking building. GTC offers passengers convenient interchange between various means of transportation. It is designed based on the principle of "prioritizing public transportation, common passengers and departure passengers". The public transportation facilities (metro, intercity railway, city bus, long distance coach, taxi) are most conveniently located to reduce the walking distance of most passengers. Concourses of metro station and intercity railway station are provided respectively in the east and west sides of GTC concourse, effectively diverting passenger flows. Parking building for private cars is right next to GTC, so passengers can reach all garage floors via a vertical transportation core.
(3) Connection to rail transit.The Baiyun Airport is connected with two underground rail lines, i.e., Metro Line 3 and an intercity railway line, both stationed on B2 in GTC. From metro and intercity railway to terminal, separate circulations are designed for departure and arrival passengers to avoid mutual interference. Besides, dedicated access is designed between the two lines of rail

transit for great efficiency and convenience of interchange.
(4) Landside transportation with separated pedestrian and vehicular circulations. A spacious pedestrian square is provided between the arrival hall on arrival floor (F1) in T2 and GTC, while the arrival bus and taxi lanes in front of the terminal are provided in the form of underground tunnels, realizing complete separation of pedestrian and vehicular circulations. Therefore, passengers can conveniently access city bus pick-up area, long-distance coach waiting area, and private car parking building on the same level or access metro station concourse, intercity railway station concourse and other floors in parking building etc. via vertical transportation.

4. Terminal Design Centering around Passenger and Baggage Flows

The design of passenger flow and baggage system process, as the basic core design for the terminal to function as a transportation building, determines the internal functional organization of the terminal and serves as a key factor affecting terminal operation efficiency.

4.1 Passenger flow design
As a large integrated terminal of a hub airport, T2 involves various passenger flows, including domestic departure, domestic arrival, international departure, international arrival, domestic-international, domestic-domestic, international-domestic and international-international transfer, international flight domestic section (domestic-Guangzhou-international, international-Guangzhou-domestic, Guangzhou-domestic- international, international-domestic-Guangzhou) etc. These flows intersect, connect or integrate into new flows, leading to a quite complicated overall situation. In the design, the strategies of mixed domestic flows, separated international flows, and swing stands are adopted to work out the internal organization and floors in the terminal.

(1) Short walking distance for major pedestrian flows. The walking distance from the curbside in the drop-off area to the farthest boarding gate is no more than 700m and requires a walking time of less than 9 minutes;
(2) One-way passenger flow. The strategy of single direction with less options is adopted in flow design. Once entering the check-in hall, passengers have only one direction to proceed and can arrive at the security check hall with no other options. After the security check, they only need to select a boarding gate. The arrival passengers, likewise, converge from different piers into the main building to pick up baggage. One exit is provided respectively for domestic and international passengers, while the shared arrival hall is connected to GTC. The concise spatial layout contributes to extremely clear orientation.
(3) Mixed domestic flows. Tthe arrival passengers, after deplaning, converge with departure passengers to baggage pick-up hall. Compared to separate design for departure and arrival flows, the mixed design for domestic flows not only reduces space for independent access and service room on the arrival floor, but also shares service facilities including retail, rest rooms, information desk etc., which in turn reduces the staff required at boarding gate, cuts the investment and enhances the utilization efficiency of resources.
(4) Separate international flows. Subject to customs and border inspection and relevant control requirements, the conventional design to separate departure flow from arrival one is adopted.
(5) Swing stand flows. International and domestic passengers can wait on different floors at the same time, with the fixed end of the boarding bridge to realize the functional connection and conversion and improve the stand efficiency.
(6) Convenient transfer design. Transfer flow, as one of the most important flows in T2, has some very unique design. The arrival, transfer and waiting for domestic-domestic and international-domestic transfer passengers are all on F2. Domestic–international and international-international transfer passengers, after converging through the transfer access in the main building, are led by vertical transportation to the international departure passenger CIQ entrance. The departure/arrival inspection site and resources are shared for the inspection of transfer passengers, which not only saves space and human resources but also reduce the complexity of signage system with less options available for passengers.
(7) Parallel international and domestic security check halls. The most prominent feature of T2 in terms of airport inspection process is to place the international and domestic security check in parallel as the first pass. This takes advantage of the staggered peak hours of international and domestic security check to realize the conversion between international and domestic flights, hence a feasible solution to congestion of flows during special peak hours (such as the Spring Festival).
(8) APM system planning for both long-distance and transfer flows. T2 features a "linear+ pier", layout, with piers far from the main building. To address this issue and realize the passenger and floor transfer in T1 and T2, two APM stations are reserved in T2, one for centralized domestic departure and arrival, and the other for international-domestic and domestic-international transfer. Passengers may take APM to conveniently access remote piers and T1 for boarding, or quickly access the baggage pick-up

OVERVIEW

process and transfer inspection flow.

4.2 Baggage system and technical coordination
The baggage system of the terminal building consists of such subsystems as baggage check-in, security inspection, automatic sorting and movement, automatic locker, special handling, and baggage conveyor etc. Based on the core process technology of automatic sorting and movement subsystem, T2 adopts the DCV automatic baggage sorting system, which is first of its kind in China and known for quick sorting, low error rate and high efficiency. The baggage room is provided on F1, the same level with apron, so that baggage vehicles can operate efficiently between the apron and the baggage room.

5. Cloud Concept to Create Original Building Shapes and Spaces

The light, floating and flowing features of clouds define the overall setting and tone of T2 design which features flowing form, light spatial appearance and gentle spatial style as detailed below.

(1) The main theme of the continuous cloud-shape arc is generated from the concept of cloud. The continuous multi-span arc tensile membrane canopy echoes with the continuous arc eave of the south facade. The diffuse reflection of the membrane gently reflects the metal texture of the eave, creating a diversely-layered rolling cloud image with the change of light and shadows.
(2) Inspired by the concept of cloud, the concise and smooth curvy shape integrates buildings of different volume on both east and west wings with the main building, finally realizing harmony with the natural environment.
(3) The concept of cloud is interpreted in a systematic and progressive manner. From building appearance to interior space, from the outside to the inside, everything is like the progressive layers of flowing water. From departure curbside to check-in hall, three corridors run through the security check hall, the Lingnan-style gardens, the dynamic core commercial area and finally to the waiting pier. From urban space outside to public space inside, the concept of cloud flows and changes throughout building forms and spaces in a rhythmic and fluent manner, shaping the original architectural space and appearance of T2.
(4) Use wedge lines and shapes for the expression of fluidity. Intersecting wedge lines and shapes are introduced in design, which are different from orthogonal shapes. Due to the illusion of unparalleled wedge lines in perspective, a sense of instability is created in space, resulting in good fluidity.
(5) Introduce natural light and shadows to create fluidity in spaces, and at the same time, provide natural guidance and comfortable experience for passengers.
(6) Create major spatial nodes with the approach to shared spaces to connect traffic and commercial spaces, which also feature fluidity.
(7) With the unique elements of Baiyun Airport retained, including the curvy main building, Y-shape column and tensile membrane canopy, T2 maintains a harmonious and consistent architectural style with T1, presenting a holistic image of the "Gemini" Terminal together with the latter.

6. Roof and Facade System to Realize Desired Building Form

6.1 Metal roofing system
The total area of metal roof is about 260,000 m^2. The non-linear arc roof is continuous and smooth, with seamless transition between the main building and the connected buildings. The project adopts outer gutter to avoid leakage and Al-Mg-Mn alloy vertical edged roof board system. The latter is comprised of Al-Mg-Mn alloy roof waterproof board, glass fiber insulation layer, flexible waterproofing layer, fiber cement board supporting layer, rock wool supporting layer, vapor barrier, profiled steel base plate supporting layer, primary and secondary purlins. This system of mature technologies can help realize the desire building form as well as functions of thermal insulation, drainage and waterproofing, daylighting etc.. It features short construction period, easy construction and maintenance, and controllable cost. Wind resistance of metal roof is reinforced through three measures. The main building roof is designed with a row of neat strip skylights, and a stripe skylight is arranged in the middle of the pier roof. The concise and practical skylight design also contributes to the aesthetics of roof design. toughened laminated insulating glass (Low-e) is used.

6.2 Facade System
T2 has a total glass facade area of about 100,000 m^2, of which the largest part is on the front facade of the main building, measuring 430m wide and 36m the highest. Different from T1 using point-supported glass facade, T2 adopts a glass facade system with visible horizontal and hidden vertical frames. The glass panel is sized 3mX2.25m, with the height division grid the same as the floor height module of 2.25m, and width division grid the same as that of planar module. The secondary structure adopts multi-level steel truss as the main load-bearing member, transverse aluminum alloy beam as wind resistance member and vertical hanger as vertical load-bearing member. Aluminum alloy beam and horizontal sunshades use integrated

extruded aluminum profile. Vertical hangers are hidden in glass joints, presenting a transparent facade unit. The multi-level steel truss structure does not need to be secured by horizontal lateral bars, and the facade has a strong sense of integrity, transparency and aesthetic beauty. MEP pipelines are integrated in the secondary structure of facade with pipelines buried in U-shape structure, and the front side covered by aluminum panels. For fire smoke exhaust, the angle of operable window should reach 60°, so pneumatic smoke exhaust window with large opening angle can be used. Considering energy conservation in Guangzhou, which is hot in summer and warm in winter, wing type adjustable electric louvers with 0°, 30°, 60° and 90° opening angles are equipped outside the exterior facade on west side. Micro-perforated panel is selected for louvers, which can provide both sunshade and visibility.

7. Integrated Architecture, Interior and Structural Design

7.1 Steel tube column+grid structure for the main building

Compared with T1, T2 has a major change in its structural system. It adopts the ordinary grid structure of mature technology. Column grid of check-in hall is 36m in horizontal direction and 53m+45m+54m in vertical direction with strengthening measures at large vertical spans, which is more cost-efficient and can meet construction schedule. Steel tube columns are adopted to reduce section size, and the two rows of columns in the hall are concealed in 12 check-in islands, so they do not interfere with the main passenger activities. Column grid in piers is 18m in vertical direction, 35m and 45m in horizontal direction. Pre-stressed concrete columns are used with tapered section to optimize the spatial effect.

7.2 The original ceiling design based on the concept of gradually spinning and changing clouds is the first of its kind in China
The interior design and dynamic architectural style are completed in a coherent manner, creating a light, floating and flowing motion compatible with the structural form. Passengers can experience open, bright and comfortable semi-outdoor space inside the building. Take the check-in hall as an example. Its ceiling is the first gradually spinning and changing ceiling in China. The rolling ceiling design with a module unit of 36m appears concise and dynamic, encompassing the strengthening grid. Aluminum alloy strips that are 3,000mm long, 500mm wide and 50mm high are installed with every 5 degree of rotation. The space under the strip skylight is made as transparent as possible, and ceiling without skylight is made flat. This not only meets skylight daylighting requirements but also avoids a completely exposed grid, forming a holistic and dynamic interface of the interior space.

7.3 Check-in island design resembling the dynamic shape of aircraft
The characteristics of aircraft is incorporated into the design of check-in islands to realize a dynamic nonlinear island shape with strong visual impact. Various disciplines, including architecture, structure, MEP, ELV, baggage, security check and security control, are coordinated to meet the integrated functions of check-in islands.

7.4 Boarding bridges with complete functions and flows in a unique shape
T2 uses swing stands and fixed ends for boarding bridges (including A380) to accommodate CSN's future demands of aircraft mix and business development. The concept of swing stands is introduced, turning the best linear stands with the most convenient circulation and the shortest flow into convertible domestic/international stands. Meanwhile, fixed ends are provided for boarding bridges to connect the stands with the arrival and departure floors in the terminal, so as to improve the operation flexibility of stands and the passenger handling efficiency while meeting the needs of the rapid business growth of the airline.
The appearance design continues the design approaches for terminals, with bold oblique cut for building shape and diagonal dividing lines for vertical aluminum plate on facade to continue the dynamic design of the terminal. This also offers smooth transition in volume from the main building, effectively reducing energy consumption due to less space and volume.
In a span of 18 to 24m, the steel structure is a giant steel truss system. Rigid connection is required between the vertical web members and upper/lower chords, while connection of the two giant steel trusses is realized via steel frame beams.

7.5 Unique tensile membrane canopy design
As an unique facility and icon of the Baiyun Airport, the tensile membrane canopy well fits the rainy and hot climate in Lingnan Region. The curbsides from T1 to T2 are all covered with such canopies to shelter passengers from wind and rain for easy pick-up or drop-off in case of unfavorable weather.

Frame membrane structure is adopted in membrane structural design with steel structure as the frame, cast steel as key nodes, and PTFE as membrane material.

8. Showcase of Distinctive Regional Culture

The Lingnan gardens on F3 and F5 allow

OVERVIEW

passengers to experience the charm of traditional gardens in a modern terminal setting. This is the best practice to integrate man and nature. Being here is like encountering an oasis during a long journey, and passengers can experience the climatic features of Guangzhou as a city in southern China. Lingnan gardens are narrow and long, with the design of oblique lines on pavement to create vivid and intriguing spaces. Coupled with tree pool, waterscape, stepping stones on water surface and other design, much can be reflected in small details and the orderly spaces. International arrival passengers from the north pier can enjoy the beautiful scenery of Lingnan gardens and make a good first impression on Guangzhou.

A "Spaceship", the culture square of check-in hall, highlights the unique charm of public culture as a special platform for displaying public culture, art and commerce, where commercial promotions can also be launched. The Lingnan garden, culture square, multi-dimensional experience of fashion and tradition, public art, unique landmark and cultural connotation, smooth passenger experience and time tunnel for passengers to experience the wonder of time travel all jointly constitute the cultural features of the airport.

9. A Non-typical Commercial Complex Interwoven with Passenger Flows

The concept of a non-typical commercial complex: based on the study of different cases of international and domestic airports and commercial centers, the concept of a unique terminal commerce, i.e., a non-typical commercial complex, is proposed. The most important aspect of the terminal is passenger flow. The meeting, parting and gathering passengers feature great mobility and short time of stay. Therefore, in order to increase non-airline revenue, after consultation with commercial consultants, a terminal commercial layout different in many aspects from that of other domestic hub airports is created: merge passenger flows and integrate commercial circulations; create duplex style atrium, cultivate commercial atmosphere; organize functional layout, control trade mix ratio; introduce ecology and greening and optimize the commercial environment. Based on the flows, ingeniously incorporate commercial facilities, so that passengers can enjoy a convenient shopping environment along the way, thus increasing non-airline revenue.

9.1 Merge passenger flows and activate commercial circulations
The specific measures are to integrate circulations as much as possible to avoid inactive space and circulations; adjust vertical transportation to reduce repetitive traffic area; closely integrate retail circulations with passenger circulations to increase the latter's accessibility to shops and eliminate shops of poor accessibility.

9.2 Coordinate various disciplines to increase the floor area
The specific measures are to coordinate architecture, structure, MEP and other disciplines; customize the design to reduce surplus space of structure and MEP etc; maximize commercial area to enhance the commercial value of the terminal.

9.3 Create duplex style atrium to foster commercial atmosphere
The specific measures are to integrate the planar functions; create, within the terminal setting, the kind of vertical commercial atrium that is commonly found with the traditional malls; improve spatial capacity to gather people, enhance the commercial atmosphere and add visible interface for advertisements.

9.4 Integrate functional layout and control trade mix
The specific measures are to, based on statistics and projections, rationally control the proportion of commerce, F&B and retails to ensure the convertibility between F&B and retail in future; optimize trade mix and layout, balance functional and commercial needs, and enhance the efficiency of commercial facilities.

9.5 Bring in greening to optimize the commercial environment
Green atrium or garden are provided at important areas for passenger flows and commercial activities to optimize indoor environmental quality, further improve the commercial value and reflect the characteristics of Lingnan architecture.

10. Green Building Design Oriented to Energy Efficiency and Environmental Protection

For terminal building with gigantic space and volume, the selection of envelop materials and equipment is key to building energy conservation. T2 adopts various green design and energy-efficient measures and has passed national three-star green building design certification, making the project the first three-star large-scale green public transportation hub complex in hot and humid region and an example of green terminal building.

10.1 Solid Thermal Environment Analysis
CFD fluid dynamics software and light environment analysis software are used to

finalize the floor plan as well as the locations and size of the operable parts in glass facade. Roof materials and structure are optimized based on energy-efficiency calculation. Dynamic energy consumption simulation software DeST and lighting energy consumption simulation software Daysim are used to analyze the energy consumption of AC, heating and lighting system in the building.

10.2 Rational Natural Daylighting
Based on spatial size and planar functions, the typical functional spaces of the project are provided with special natural daylighting design. Skylights are evenly planned in the roof above the check-in hall of the terminal, cutting the annual power consumption by about 2.8 million KWH, i.e. about 20% of the total lighting consumption. Daylighting wells are provided in basement garage of GTC to improve the daylighting in underground spaces.

10.3 Organization of Natural Ventilation
Average annual wind speed in Guangzhou is about 2m/s, so wind power resources are not abundant. CFD fluid dynamics software is used to guide and optimize the floor plan as well as the location and size of operable parts in the glass facade, so as to enhance natural ventilation capacity of the building and reduce AC energy consumption in summer. For areas with poor ventilation such as the internal CIQ area in the terminal, atrium gardens where passengers can take a rest are provided to improve the natural ventilation of the area, increase the building energy efficiency and upgrade the quality of indoor air.

10.4 Efficient Solar Energy Utilization
With large solar elevation angle in Guangzhou, total solar radiation and sunshine hours are

sufficient. Sufficient solar radiation and sunlight offer favorable conditions for the utilization of solar energy. Solar-powered heat pumps are used in the hourly-rate hotel, first-class and business-class cabins of T2, accounting for 72% of domestic hot water consumption. Solar PV power panels are used on the metal roof above the security check hall of the terminal.

10.5 Energy-efficient and Comfortable Sunshade Design
Wing-type adjustable electric louvers are used for the west side of T2, while indoor electric roller shutters are installed in the east side.

Conclusion

Architecture design of mega terminal is a typical example of multidisciplinary integration and innovation. Architects need to gain a deeper understanding of the user needs and work closely with consultants of other disciplines.

Besides, they must also integrate cross-disciplinary and sophisticated technical systems, and, on this basis, make great efforts in innovations and conduct collaborative and integrated design across disciplines and specialties. Only by doing so could it be possible to realize the grand vision of creating a gateway aviation hub. Since its operation, T2 has been widely acclaimed by the industry and the general public. Skytrax recognizes T2 as a 5-star terminal and has given Baiyun Airport such awards as World's Most Improved Airline and Best Regional Airport - China. Guangzhou Baiyun International Airport, as the headquarters base of China Southern Airlines, serves as an air bridge to connect Guangzhou, known as China's southern gateway, with Europe, the United States, Australia and Africa. The newly built T2 well complementing with T1 will make its due contribution to the grand strategic goal of making Guangzhou a core city in GBA and China a leader in world's civil aviation industry.

第2章

图纸
CHAPTER 2 **DRAWINGS**

2.1　效果图
2.2　总平面图
2.3　各层平面图
2.4　立面图、剖面图

2.1　RENDERING
2.2　MASTER PLAN
2.3　FLOOR PLANS
2.4　ELEVATIONS AND SECTIONS

2.1 效果图　　　　　　　　2.1 RENDERING

广州白云国际机场鸟瞰效果图
Bird's Eye View of Baiyun Airport

二号航站楼本期工程鸟瞰效果图
Bird's Eye View of Current Phase of T2

▼ 出发车道边鸟瞰效果图
Bird's Eye View of Departure Curbside

出发车道边低点透视效果图
Low-viewpoint Perspective of Departure Curbside

主楼与东五指廊交接位置效果图
Rendering of Connection between Main Building and East 5th Pier

交通中心与航站楼半鸟瞰效果图
Mid-air Bird's Eye View of GTC and T2

办票大厅效果图
Rendering of Check-in Hall

▼ 迎客大厅及办票大厅效果图
Rendering of Arrival Hall and Check-in Hall

岭南花园效果图
Rendering of Lingnan Gardens

东连接指廊效果图
Rendering of East Connecting Building

西五指廊效果图
Rendering of West 5th Pier

2.2 总平面图　　2.2 MASTER PLAN

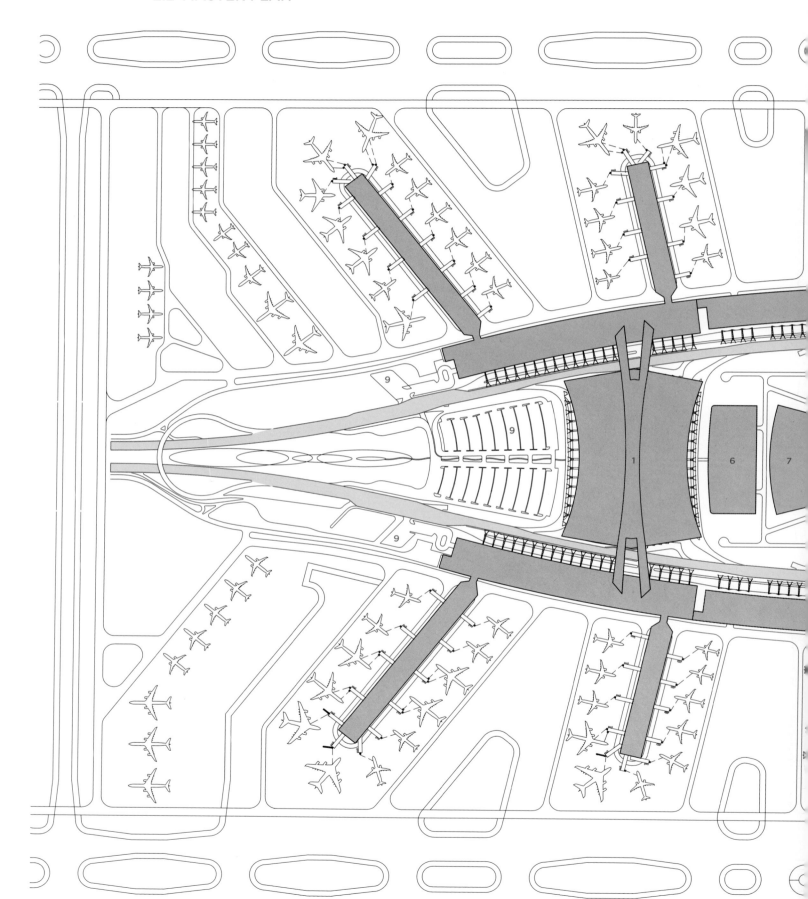

总平面图
Master Plan

1. 一号航站楼
2. 二号航站楼
3. 交通中心停车楼
4. 东四指廊（未建）
5. 西四指廊（未建）
6. 停车楼
7. 机场酒店
8. 航管楼及塔台
9. 停车场
10. 空侧机坪

1. T1
2. T2
3. GTC Parking Building
4. East 4th Pier (Unbuilt)
5. West 4th Pier (Unbuilt)
6. Parking Building
7. Airport Hotel
8. ATC Building and Tower
9. Parking Lot
10. Airside Apron

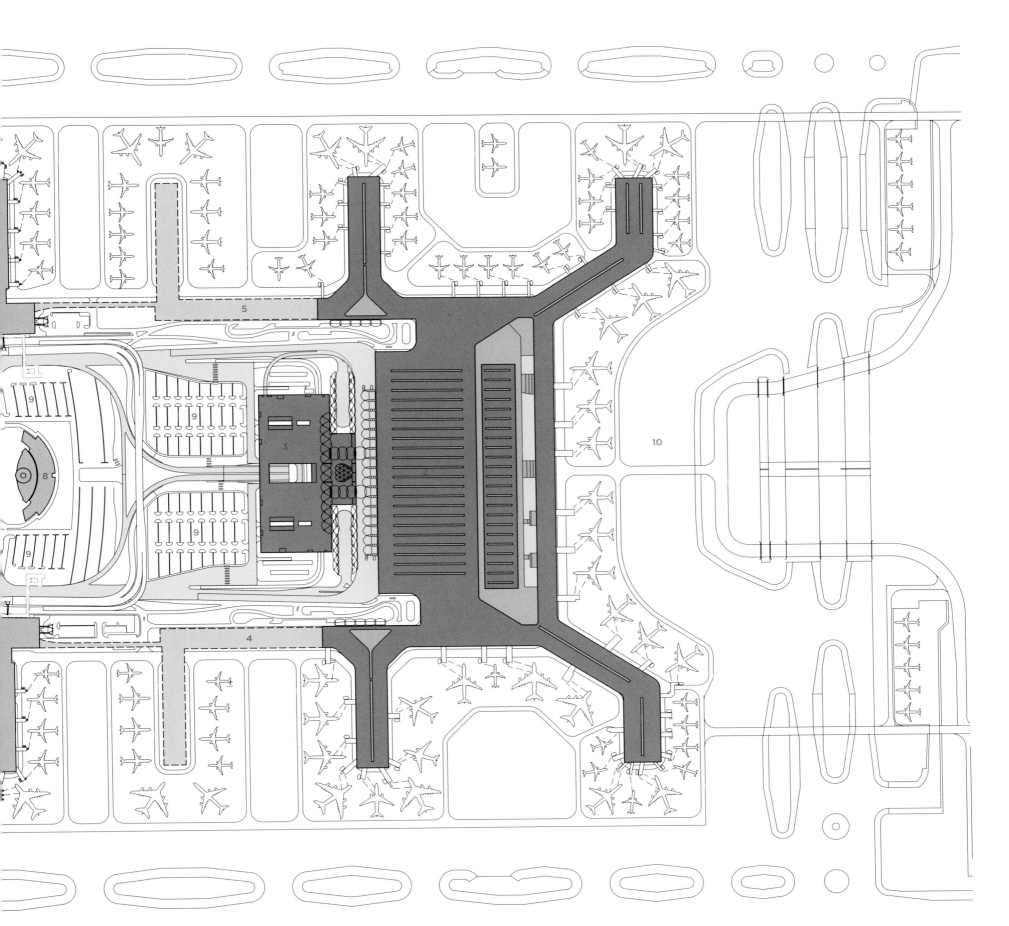

2.3 各层平面图   2.3 FLOOR PLANS

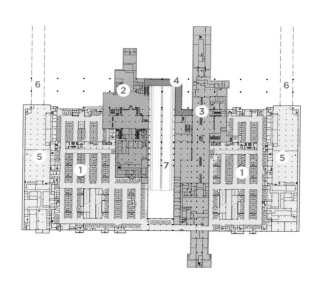

二号航站楼及交通中心负二层平面图
B2 Plan of T2 and GTC

1. 停车区域
2. 地铁站厅
3. 城轨站厅
4. 地铁、城轨互换通道
5. 设备中心
6. 设备管廊
7. 北进场隧道

1. Parking Area
2. Metro Station Concourse
3. Intercity Railway Station Concourse
4. Metro-Intercity Railway Interchange Passage
5. MEP Center
6. MEP Tunnel
7. North Approach Tunnel

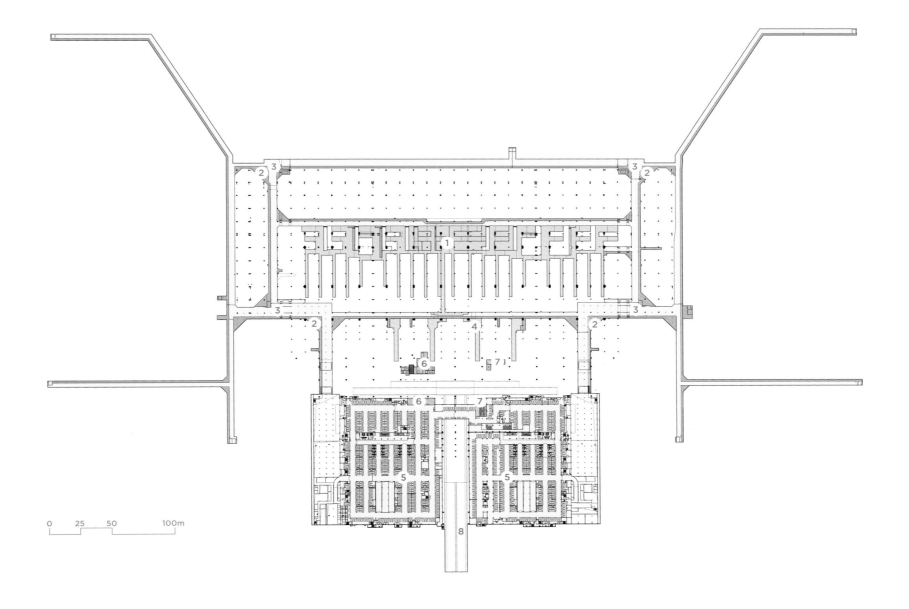

二号航站楼及交通中心负一层平面图
B1 Plan of T2 and GTC

1. 行李系统管廊
2. 电气管廊
3. 空调、水专业管廊
4. 通风管廊
5. 停车区域
6. 地铁垂直交通
7. 城轨垂直交通
8. 北进场隧道

1. Baggage System Utility Tunnel
2. Electrical Utility Tunnel
3. AC & Plumbing Utility Tunnel
4. Ventilation Utility Tunnel
5. Parking Area
6. Metro Vertical Transportation
7. Intercity Railway Vertical Transportation
8. North Approach Tunnel

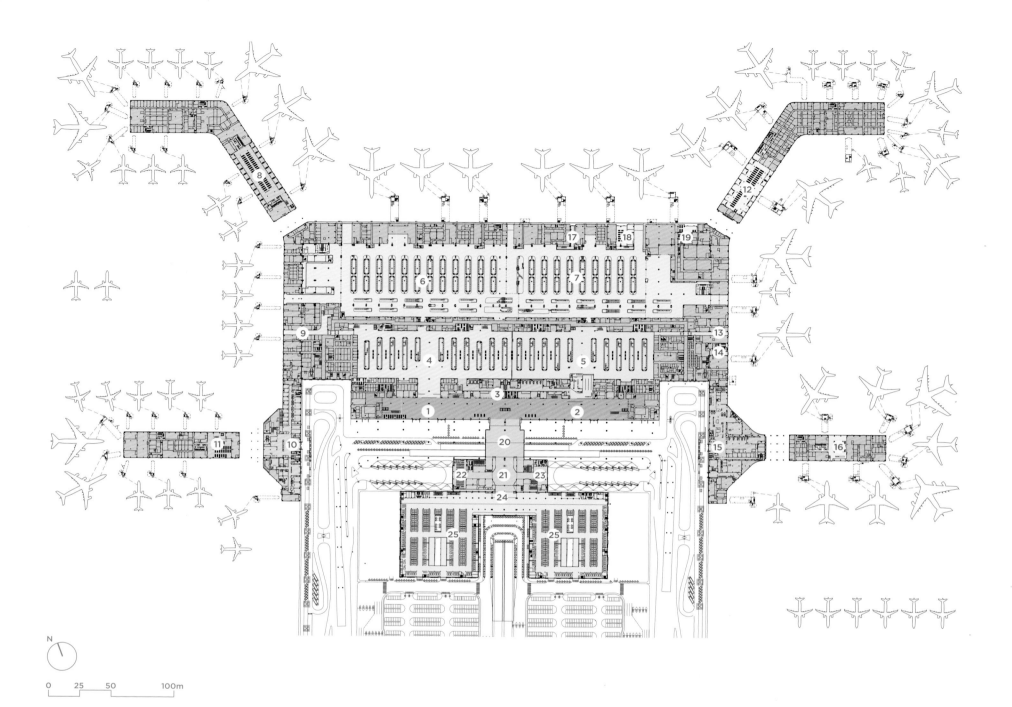

二号航站楼及交通中心一层平面图
F1 Plan of T2 and GTC

1. 国内迎客大厅
2. 国际迎客大厅
3. 商业
4. 国内行李提取厅
5. 国际行李提取厅
6. 国内行李分拣机房
7. 国际行李分拣机房
8. 国内远机位候机厅
9. 国内远机位到达走廊
10. 国内贵宾室
11. 国内误机旅客候机厅
12. 国际远机位候机厅
13. 国际-国内-广州旅客到达走廊
14. 国际远机位到达厅
15. 国际贵宾室
16. 国际误机旅客候机厅
17. 国内-广州-国际候机厅
18. 广州-国内-国际候机厅
19. 国际-广州-国内候机厅
20. 交通中心过厅
21. 交通中心旅客大厅
22. 大巴客运站
23. 旅游、员工、航延大巴
24. 私家车车道边
25. 停车区域

1. Domestic Arrival Hall
2. International Arrival Hall
3. Commerce
4. Domestic Baggage Claim Hall
5. International Baggage Claim Hall
6. Domestic Baggage Sorting Machine Room
7. International Baggage Sorting Machine Room
8. Domestic Remote Stand Waiting Hall
9. Domestic Remote Stand Arrival Channel
10. Domestic VIP Room
11. Domestic No-show Waiting Hall
12. International Remote Stand Waiting Hall
13. International-Domestic-Guangzhou Arrival Channel
14. International Remote Stand Arrival Hall
15. International VIP Room
16. International No-show Waiting Hall
17. Domestic-Guangzhou-International Waiting Hall
18. Guangzhou-Domestic-International Waiting Hall
19. International-Guangzhou-Domestic Waiting Hall
20. GTC Hallway
21. GTC Passenger Hall
22. Coach Terminal
23. Coach for Tourists, Staff and Delayed Passengers
24. Private Car Curbside
25. Parking Area

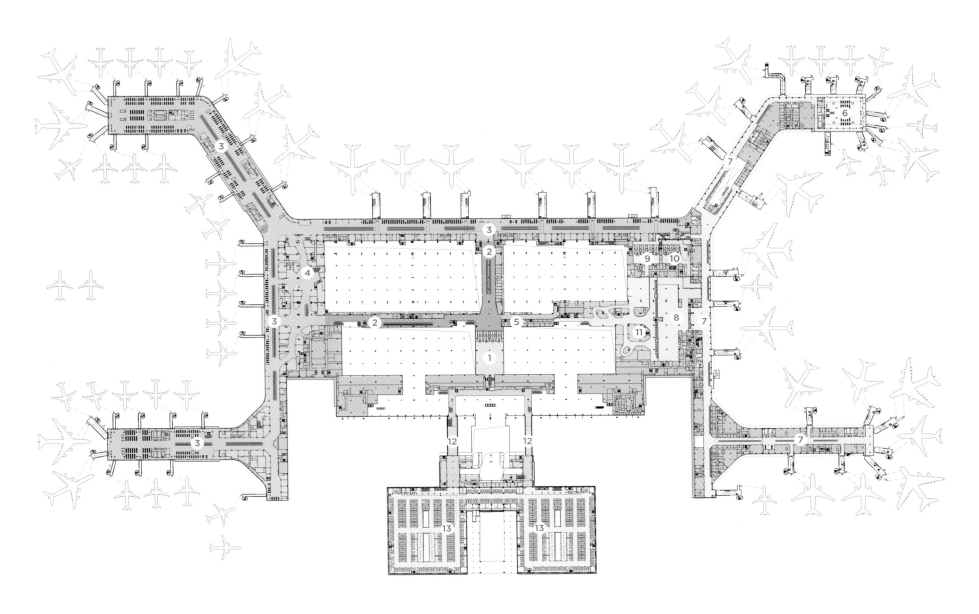

二号航站楼二层及交通中心二层平面图
F2 Plan of T2 and GTC

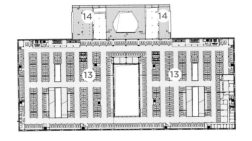

交通中心三层平面图
F3 Plan GTC

1. 中转安检候检区	1. Transfer Security Check Waiting Area
2. 国内到达走廊	2. Domestic Arrival Channel
3. 国内混流候机厅	3. Domestic Mixed-Flow Waiting Hall
4. 国内集中商业区	4. Domestic Centralized Commercial Area
5. 国内转国际区域	5. Domestic-International Transfer Area
6. 国际出发候机厅	6. International Departure Waiting Hall
7. 国际到达走廊	7. International Arrival Channel
8. 入境边检候检厅	8. Arrival Frontier Inspection Waiting Hall
9. 国际转国内区域	9. International-Domestic Transfer Area
10. 国际转国际区域	10. International-International Transfer Area
11. 国际免税商业区	11. International Duty-Free Commercial Area
12. 旅客通道	12. Passenger Channel
13. 停车区域	13. Parking Area
14. 屋面绿化	14. Roof Greening

GUANGZHOU BAIYUN INTERNATIONAL AIRPORT TERMINAL 2 AND SUPPORTING FACILITIES _ 049

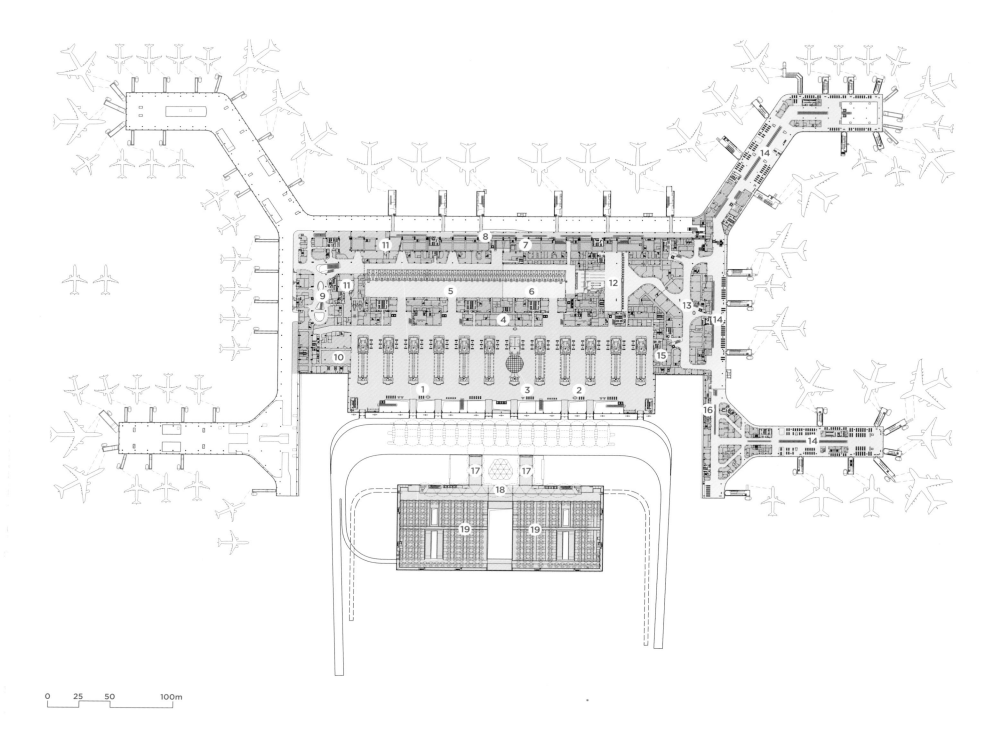

二号航站楼及交通中心三层平面图
F3 Plan of T2 and GTC

1. 国内办票大厅
2. 国际办票大厅
3. 文化广场
4. 陆侧商业
5. 国内安检大厅
6. 国际安检大厅
7. 岭南花园
8. 国际到达走廊
9. 国内集中空侧商业区
10. 国内陆侧两舱候机室
11. 国内空侧两舱候机室
12. 国际联检大厅
13. 国际集中免税商业区
14. 国际出发候机厅
15. 国际两舱值机等候区
16. 国际两舱出发通道
17. 旅客通道
18. 预留出发车道边
19. 屋面绿化停车场

1. Domestic Check-in Hall
2. International Check-in Hall
3. Cultural Square
4. Landside Commercial Area
5. Domestic Security Check Hall
6. International Security Check Hall
7. Lingnan Gardens
8. International Arrival Channel
9. Domestic Centralized Airside Commercial Area
10. Domestic Landside First/Business-Class Waiting Room
11. Domestic Airside First/Business-Class Waiting Room
12. International CIQ Hall
13. International Centralized Duty-Free Commercial Area
14. International Departure Waiting Hall
15. International First/Business-Class Check-In Waiting Area
16. International First/Business-Class Departure Channel
17. Passenger Channel
18. Reserved Departure Curbside
19. Green Roof Parking

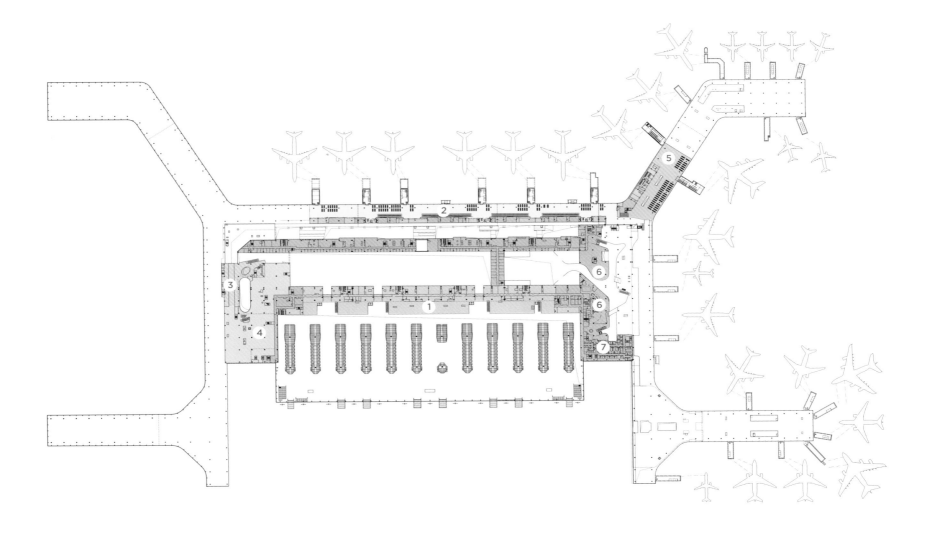

二号航站楼及交通中心四层平面图
F4 Plan of T2 and GTC

1. 陆侧餐饮平台
2. 国际出发候机厅
3. 预留APM站
4. 预留餐饮平台
5. 国内混流候机厅
6. 国际两舱候机室
7. 国际计时旅客休息区

1. Landside F&B Platform
2. International Departure Waiting Hall
3. Reserved APM Station
4. Reserved F&B Platform
5. Domestic Mixed-Flow Waiting Hall
6. International First/Business-Class Waiting Room
7. International Hourly Charged Lounge

2.4 立面图、剖面图   2.4 ELEVATIONS AND SECTIONS

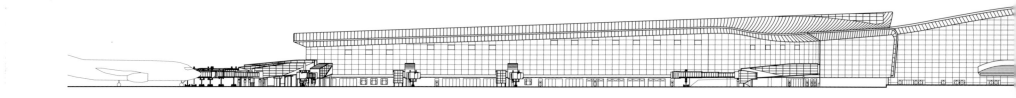

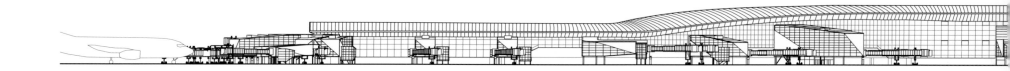

南立面图
South Elevation

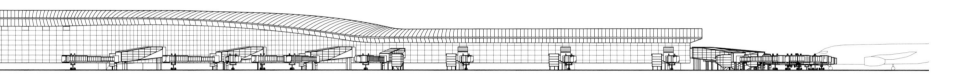

北立面图
North Elevation

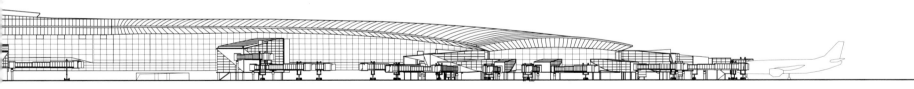

东立面图
East Elevation

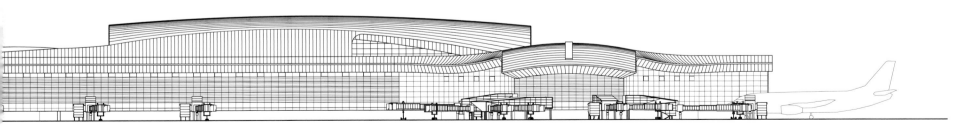

西立面图
West Elevation

GUANGZHOU BAIYUN INTERNATIONAL AIRPORT TERMINAL 2 AND SUPPORTING FACILITIES _ 054

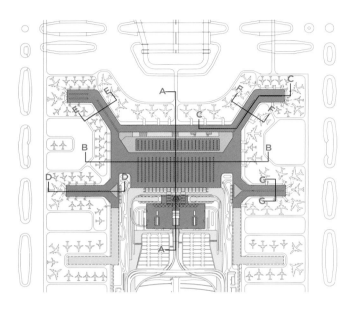

剖切位置索引
Section Position Index

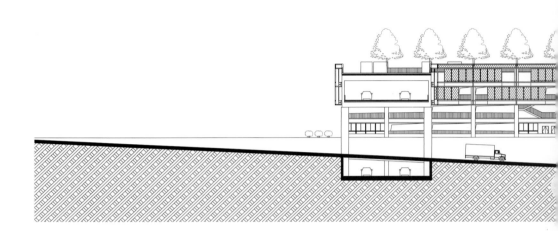

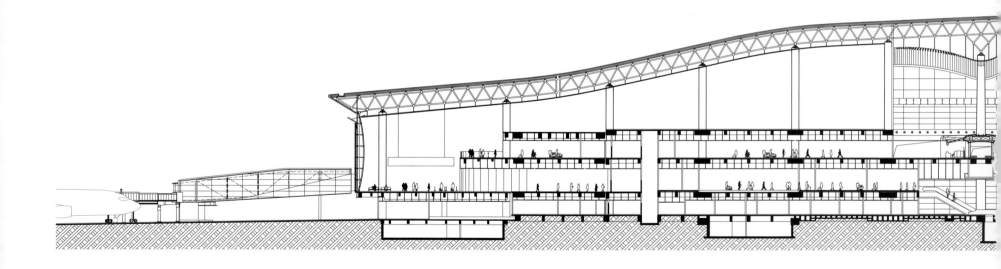

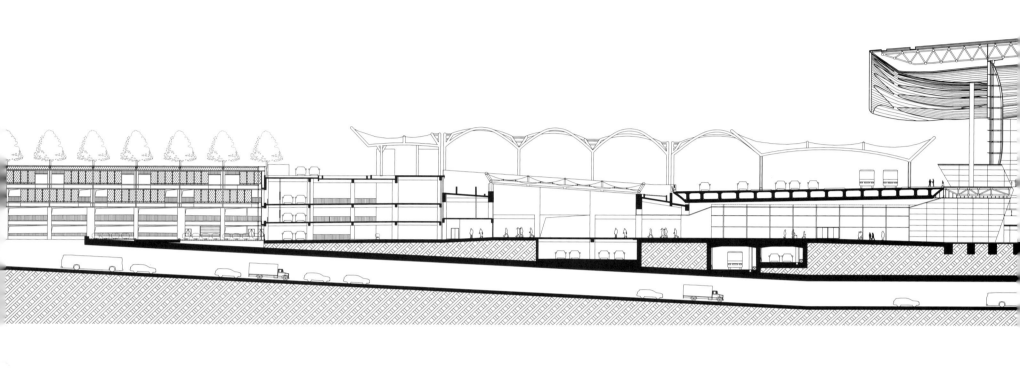
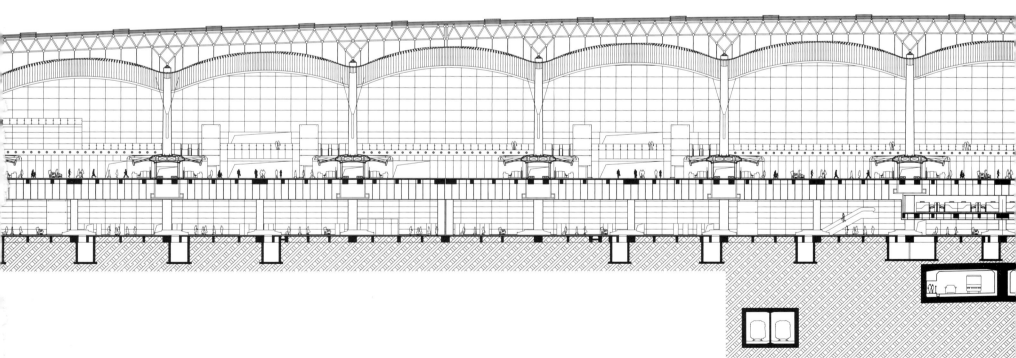

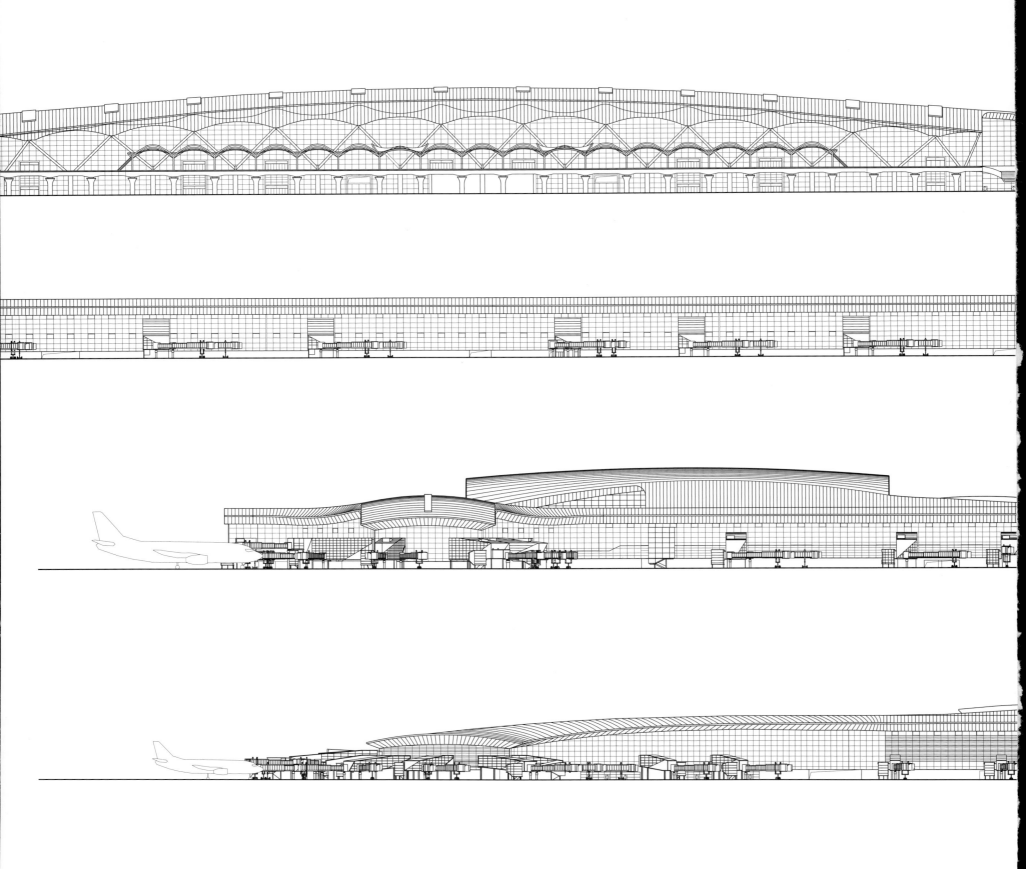

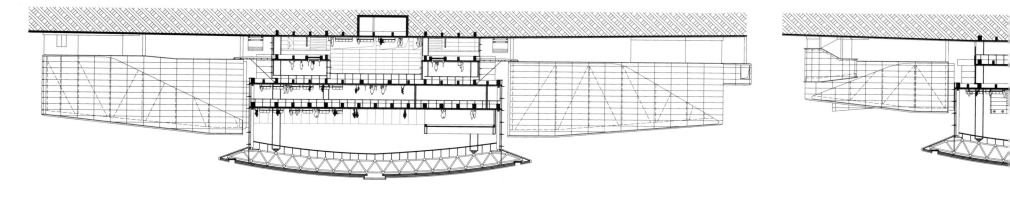
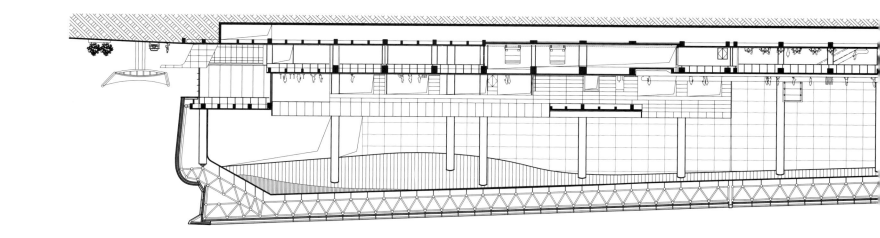
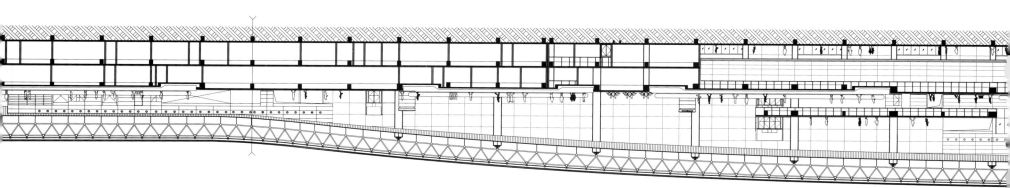

G-G 东五桥海横剖面图
G-G Horizontal Section of East 5th Pier

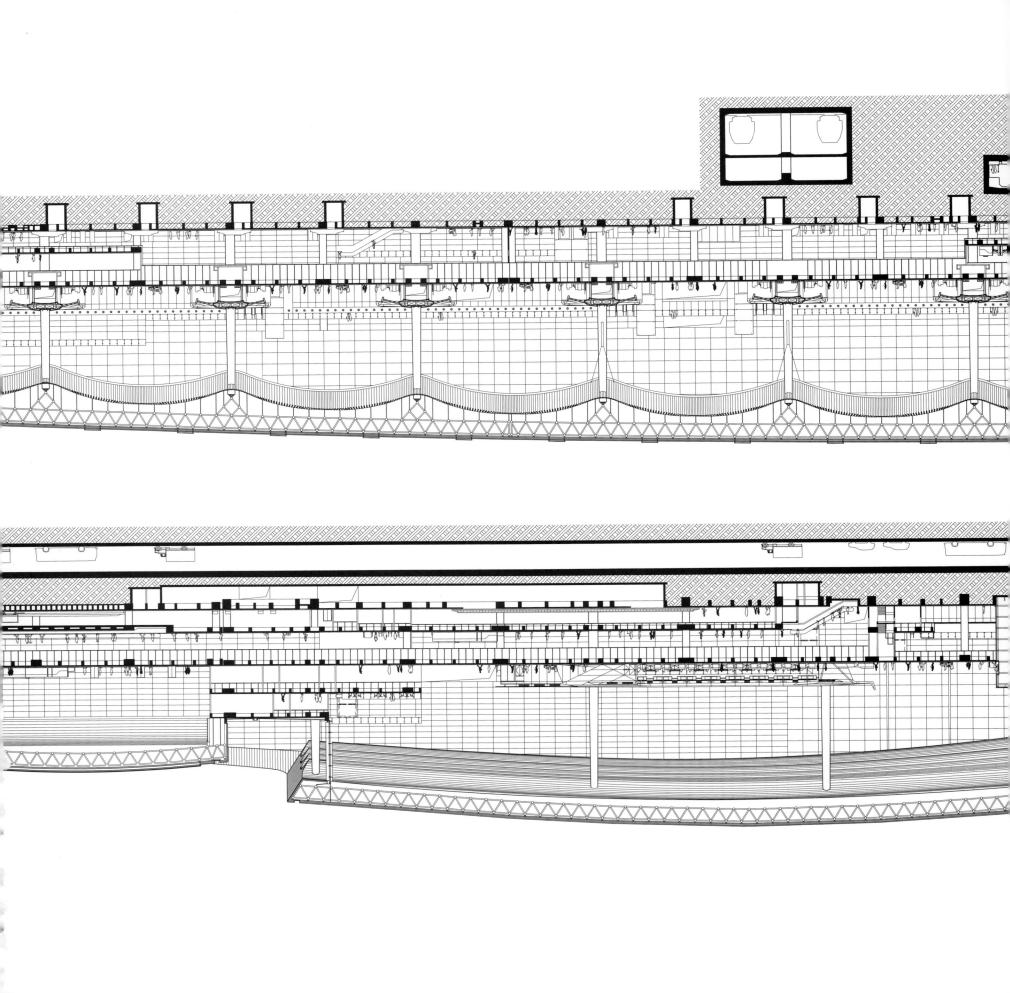

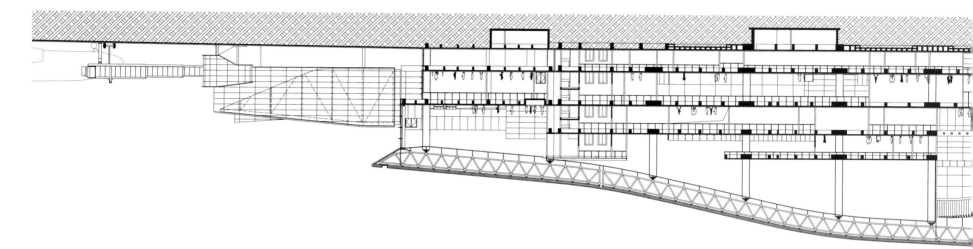

B-B 二号航站楼东西向剖面图
B-B East-West Section of T2

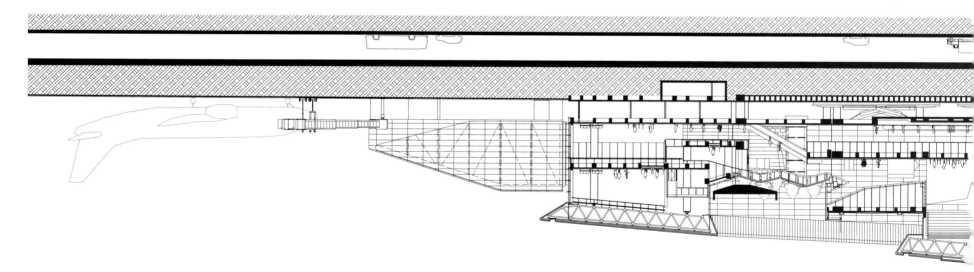

A-A 二号航站楼及交通中心南北向剖面图
A-A South-North Section of T2 and GTC

F-F 东六指廊横剖面图
F-F Horizontal Section of East 6th Pier

E-E 西六指廊横剖面图
E-E Horizontal Section of West 6th Pier

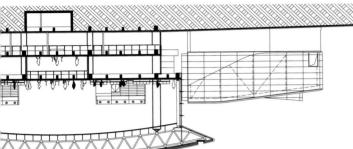

D-D 西五指廊横剖面图
D-D Horizontal Section of West 5th Pier

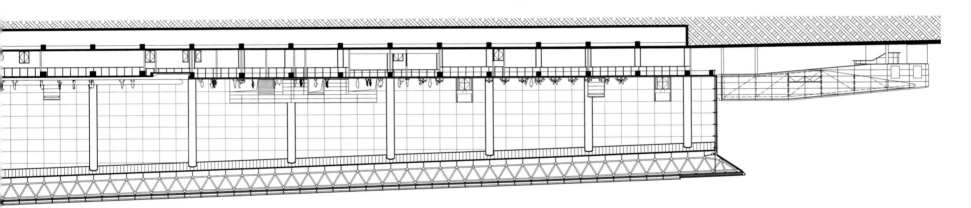

C-C 东六指廊纵剖面图
C-C Vertical Section of East 6th Pier

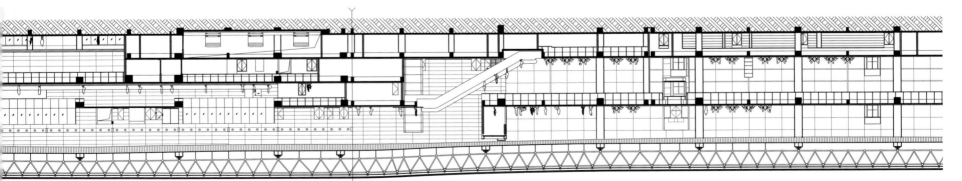

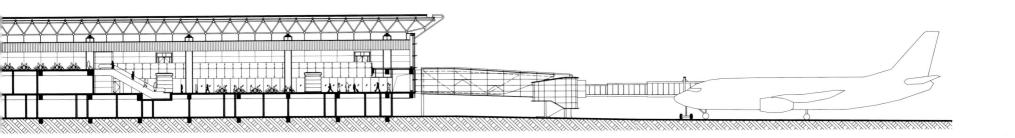

第3章
工程概述
CHAPTER 3 PROJECT PROFILE

3.1 机场定位
3.2 航站区现状及发展预测
3.3 机场总体规划与本期扩建工程
3.4 工程设计亮点

3.1 AIRPORT POSITIONING
3.2 EXISTING CONDITIONS AND DEVELOPMENT PROJECTION
3.3 MASTER PLAN AND EXPANSION PROJECT
3.4 DESIGN HIGHLIGHTS

3.1 机场定位

3.1.1 定位及其重要性

广州白云国际机场是全国三大国际枢纽机场之一，是粤港澳大湾区航空枢纽最重要的组成部分，是广东省"5+4"机场群的核心。白云机场扩建工程（二号航站楼及配套设施）是"十二五"规划和《珠三角规划纲要》重点建设项目。目前，白云机场航线网络连接欧美澳非、覆盖东南亚、辐射中国内地各主要城市，是中国南方航空的总部基地机场。2018年广州白云国际机场安全起降航班接近48万架次，全年旅客吞吐量接近7000万人次，完成货邮接近190万吨，以上排名均居全球前列。

3.1.2 扩建的必要性

1. 适应区域经济社会发展的需要。"十一五"规划指出广州市要实现经济社会发展和城市建设水平的全面提升，现代化大都市建设取得重要的阶段性进展。这样的经济战略也将为机场的发展提供重要支持，而经济与社会的发展也依托于民航的发展。

2. 满足航空业务量需求持续快速增长的需要。2009年白云机场共完成飞机起降30.8万架次，年旅客吞吐量3705万人次，货邮吞吐量95.5万吨，全面超过了一期工程年旅客吞吐量2500万人次的设计容量，许多设施已经饱和。随着近年民航旅客运输量所占运输市场份额的不断上升，为解决现有航站区和飞行区设施能力不足的突出矛盾，需要对白云机场进行扩建。

3. 建设广州枢纽机场的需要。广州白云国际机场是全国三大枢纽机场之一，白云机场的扩建不仅对广州建成国际大都市具有重要意义，而且对粤港澳大湾区经济协调发展具有重要意义。白云机场的长远发展将着眼于构建完善的国内国际航线网络，成为中国连接世界各地的空中门户、亚太地区的核心枢纽，最终成为世界航空网络的重要节点。

4. 应对亚太地区航空枢纽机场竞争的需要。白云机场建设成为亚太地区航空枢纽体现了国家战略和意志。广州机场将会同香港机场、深圳机场共同组成优势互补的珠三角区域机场体系，同周边新加坡、曼谷、成田、仁川、吉隆坡等大型枢纽机场之间就亚太地区航空运输中心的地位展开竞争，因此对广州机场基础设施规模的扩大与提升提出了迫切的要求。

3.2 航站区现状及发展预测

3.2.1 现状

一号航站楼及二号航站楼为第一航站区，设计目标年为2020年。在第一航站区东面规划第二航站区三号航站楼及三个卫星厅，设计目标年为2035年。此外，在西航站区以北进场路的西侧区域规划机场货运区及民航快递区，东侧区域规划机场机务维修区，第三条跑道以东的东航站区北侧规划为货运代理区及生产辅助设施区。

3.2.2 发展预测

为满足广州白云国际机场枢纽建设升级需求，第一、第二航站区及各航站楼均以各联盟航空公司一体化运营为分配原则。其中星空联盟、寰宇一家等成员航空公司在一号航站楼运营，天合联盟成员航空公司在二号航站楼运营，其他航空公司在三号航站楼运营。通过航空公司板块区域化划分，同板块航空公司在同一航站楼内运营，有利于提升航站区的运行效率，提高旅客服务品质。

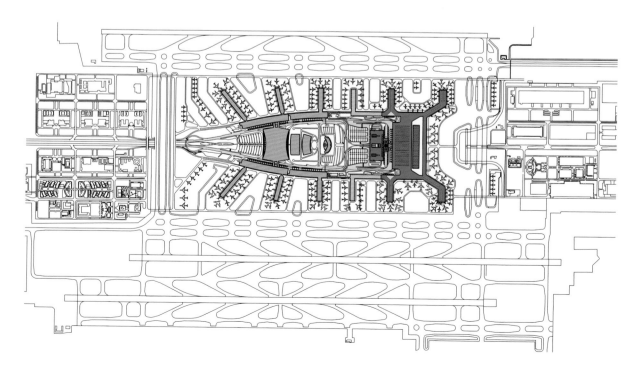

航站区现状总平面图
Existing Master Plan of Terminal Area

3.1 AIRPORT POSITIONING

3.1.1 Positioning and Significance

Guangzhou Baiyun International Airport (Baiyun Airport) is one of the three major international hub airports in China, a significant part of the air transport hub of Guangdong-Hong Kong-Macao Greater Bay Area (GBA), and the core of the "5+4" airport cluster of Guangdong. Baiyun Airport Expansion Project (Terminal 2 and Supporting Facilities) is a key construction stipulated in the Twelfth Five-year Plan and the *Outlines of the Plan for the Pearl River Delta*. As the headquarters base airport of China Southern Airlines (CSZ), Baiyun Airport holds air routes connected to Europe, Northern America, Australia, Africa, Southeast Asia and major cities in China. In 2018, Baiyun Airport served nearly 70 million passengers and 1.9 million tons of freight & mail with 480,000 aircraft movements, all among the highest in the world.

3.1.2 Necessity of Expansion

1. To adapt to regional economic and social development. As per the Eleventh Five-year Plan, Guangzhou will level up its economic and social development as well as urban construction in full scale, and make key staged headways in becoming a modernized metropolis. Such economic strategy will significantly support the development of the Airport, while economic and social development likewise relies on the development of civil aviation.

2. To keep up with the ever-growing aviation business. In 2009, Baiyun Airport registered 308,000 aircraft movements, with annual passenger traffic of 37.05 million and freight & mail traffic of 955,000 tons. That exceeded the designed annual passenger capacity of 25 million for Phase I, saturating many of its facilities. Given the constantly increasing volume of air passengers, the then insufficient capacity of the terminal area and movement area called for an expansion.

3. To support Guangzhou's hub airport strategy. Baiyun Airport is one of China's three major hub airports. Its expansion is of great significance to build Guangzhou into an international metropolis and ensure coordinated economic development of GBA. In the long term, Baiyun Airport will focus on establishing a sound network of domestic and international airlines, serving as an air gateway for China to connect with the world, a core hub in the Asia-Pacific region, and eventually an important node of the global aviation network.

4. To cope with the competition among hub airports in the Asia-Pacific region. It is a national strategy to develop Baiyun Airport into an air transport hub of the Asia-Pacific region. It will work together with Hong Kong International Airport and Shenzhen Bao'an International Airport to form a complementary airport system in the Pearl River Delta to compete with peripheral large hub airports in Singapore, Bangkok, Narita, Incheon and Kuala Lumpur. This also urges the expansion and upgrading of infrastructure in Baiyun Airport.

3.2 EXISTING CONDITIONS AND DEVELOPMENT PROJECTION

3.2.1 Existing Conditions

Terminal Area 1 consists of Terminal 1(T1) and Terminal 2 (T2), whose design will be finished by 2020. The design of Terminal Area 2, including Terminal 3 (T3) and three satellite halls, planned on the east of Terminal Area 1, is expected to finish by 2035. It is planned to provide freight area and air express area on the west side, and aircraft maintenance area on the east side, of the approach road in the north of West Terminal Area. It is planned to provide freight forwarding area and production auxiliary facilities area in the north of East Terminal Area on the east of the third runway.

3.2.2 Development Projection

To support the upgrading of Baiyun Airport, Terminal Area 1 and 2 and all terminals are designed on the principle of ensuring integrated operation of each aviation alliance. Specifically, airline members of Star Alliance and Oneworld will operate in T1, those of SkyTeam in T2, and other airlines in T3. Placing airlines of the same alliance in the same terminal will greatly improve the operation efficiency of the terminal areas as well as service quality for passengers.

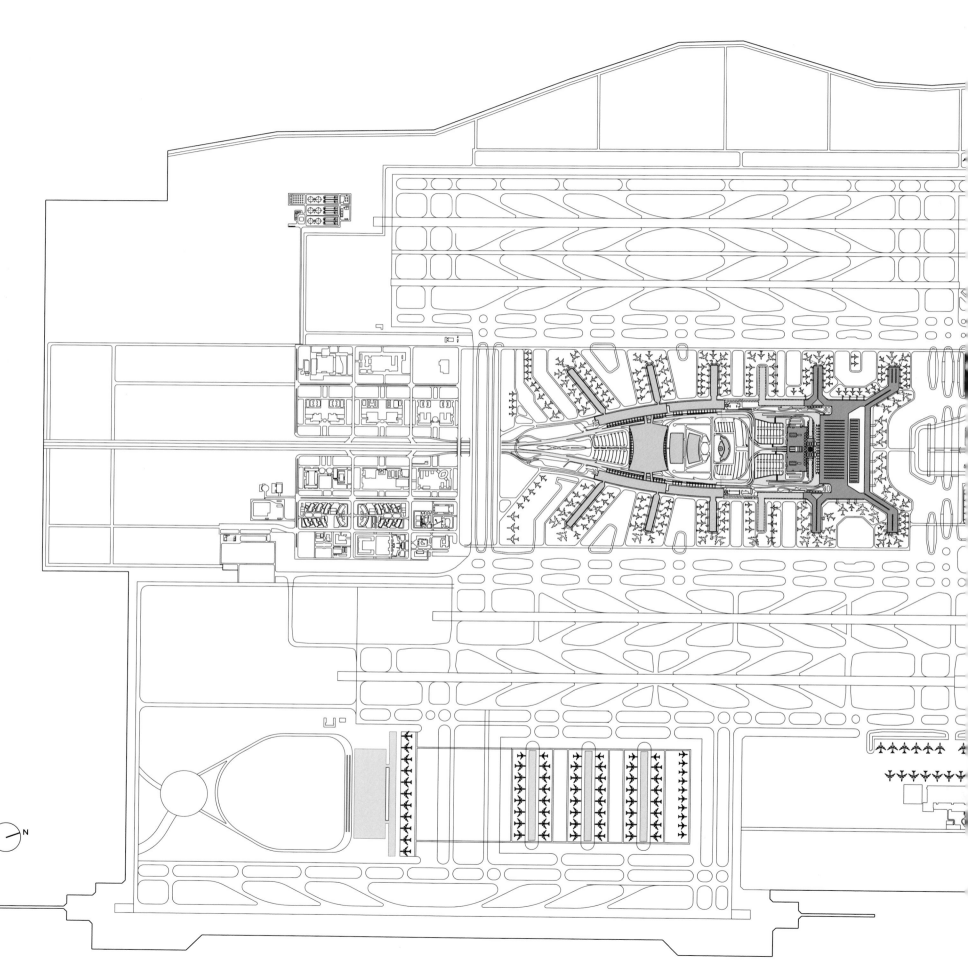

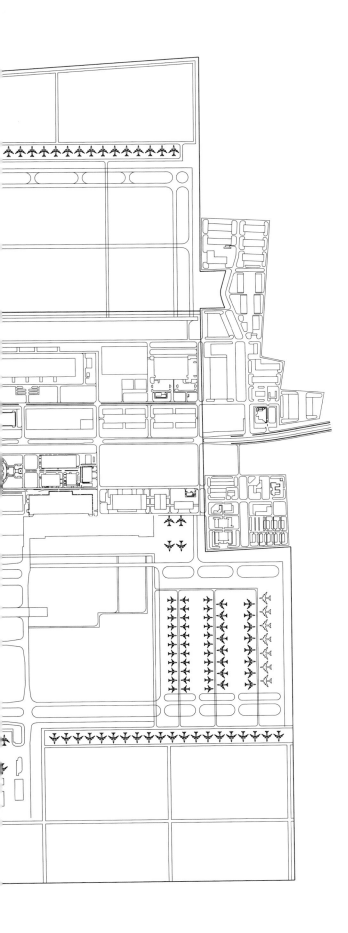

航站区总体规划平面图
Master Plan of Terminal Area

3.3 机场总体规划与本期扩建工程

3.3.1 航空业务量预测与建设规模

1. 机场参数预测。一号航站楼于2004年8月5日正式投入使用，同时机场开始了机场中远期发展规划的进一步研究，启动总体规划修编工作。机场总体规划修编于2007年9月由民航总局批复。在2007版总体规划基础上，中国民航机场建设集团公司编制了《广州白云国际机场扩建工程项目建议书》，扩建工程以2020年作为设计目标年，预测机场旅客吞吐量为8000万人次，两个航站楼的比例分配为一号航站楼3500万人次，二号航站楼4500万人次。货邮吞吐量250万吨，飞行起降量62万架次。

2. 二号航站楼建设规模预测。根据《关于广州白云国际机场扩建工程（可研报告）的咨询评估专家组意见》二号航站楼的设计容量由3000万人次调整为3500万人次（国内航班使用），二号航站楼的设计容量为4500万人次。结合民用机场工程项目建设标准（建标105-2008）提出的航站楼建筑面积指标，结合超大型机场综合管沟、登机桥固定端、机电设备用房、站坪架空层等特有设施及为未来发展预留各类旅客服务设施、机电设备及捷运系统等使用空间等因素，推导出以每百万旅客对应1.46万平方米作为规模控制的参考指标，二号航站楼总建筑面积（功能面积+非功能面积）为65.87万平方米。

3.3.2 航站区总图工程

1. 陆侧地面交通系统工程。陆侧交通系统设计应能够满足二号航站楼旅客吞吐量4500万人次的陆侧交通需求，并达到较高的服务水平。建设规模及标准应与交通需求、道路功能定位相匹配，实现适用性和经济性最佳结合。二号航站楼陆侧交通系统工程主要包括北进场隧道、东环路、出港高架桥、西环路、南往南高架桥等28条道路，交通中心南面北进场隧道东西两侧设置了两个地面停车场，地面停车场总面积53089平方米，共有私家车位1528个。

2. 陆侧桥梁工程。陆侧市政桥梁工程共包括主线桥1座，匝道桥5座，分别为出港高架桥、GTC私家车坡道、南往南匝道、北进场东匝道、北进场西匝道、东三匝道和西三匝道。

3. 陆侧通信管网工程。北进场隧道属于二类隧道，大巴隧道及的士隧道属于四类隧道。设备安装内容主要包含配电系统、电气照明系统、建筑物防雷与接地、消防及弱电系统。

4. 空侧站坪与滑行道。空侧站坪工程包括二号航站楼本期远、近机位站坪的新建以及相应滑行道系统的改造。东五指廊与东六指廊间设置两条F类滑行通道。西四过夜机位与西五指廊间设置两条C类机位滑行通道，西五指廊与西六指廊间设置四条C类机位滑行通道，东六指廊、西六指廊与北垂滑之间各设置一条C类滑行通道和一条E类滑行通道，主楼北侧站坪设置两条E类机位滑行通道。

3.3.3 二号航站楼工程

1. 二号航站楼。二号航站楼是超大型国际枢纽航站楼，位于一号航站楼的北面，设计年旅客量4500万人次，本期扩建总机位78个，其中近机位64个，建筑面积65.87万平方米，为四层混凝土大跨度钢结构建筑，建筑高度约44.5m。

2. 交通中心及停车楼。交通中心及停车楼位于二号航站楼主楼的南面，总建筑面积20.84万平方米，建筑高度约12.95m。为地下二层、地上三层的钢筋混凝土框架结构建筑，实现陆侧各类交通方式有机换乘和旅客便捷停车。

3. 旅客捷运系统预留。根据IATA的C类服务标准，步行距离超过750m需要设置捷运系统。二号航站楼入口中心至最远端指廊为690m，其在本期建设中预留旅客捷运系统（APM）的空间及结构荷载，为将来出发、到达特别是中转旅客提供高水平服务。为与一号航站楼构建成便捷的一体化航空港枢纽提供必要条件。

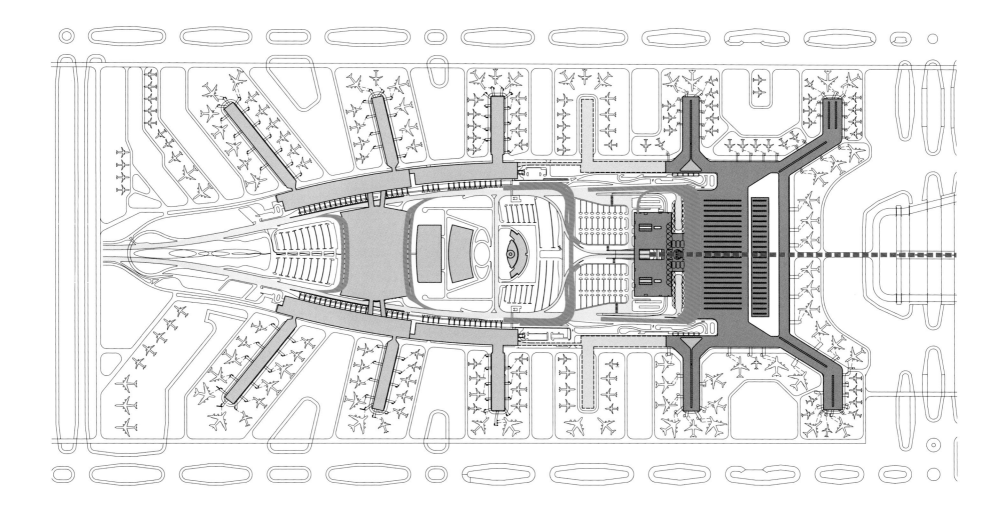

陆侧高架桥及隧道
Landside Viaduct and Tunnel

陆侧高架桥
Landside Viaduct

北进场隧道
North Approach Tunnel

3.3 MASTER PLAN AND EXPANSION PROJECT

3.3.1 Projections of Aviation Portfolio and Construction Size

1. Airport parameters. T1 was put into service on August 5, 2004. At the same time, the Airport started further study for its mid-term and long-term development plan and initiated the modification of its master plan, which was approved by the Civil Aviation Administration of China in September 2007. Based on the 2007 version of master plan, China Airport Construction Group Corporation formulated the *Proposal for the Expansion Project of Guangzhou Baiyun International Airport*. The Proposal expects Baiyun Airport to have passenger traffic of 80 million (35 million for T1 and 45 million for T2), freight and mail traffic of 2.5 million tons and 620,000 aircraft movements by 2020.

2. Construction size of T2. As per the *Opinions of the Consulting and Assessing Panel on the Expansion Project of Guangzhou Baiyun International Airport (Feasibility Study Report)*, the design capacity of T2 was adjusted from 30 million to 35 million passengers for domestic flights and 45 million passengers in total. Based on terminal floor area indices stipulated in Construction Standards of Civil Airports (Jian Biao 105-2008), as well as special facilities in super large airports such as utility tunnel, fixed ends of boarding bridges, MEP rooms and open-up floor of apron, and reserved passenger service facilities, MEP rooms and people mover system for future development, a reference index of 14,600 m² for every 1 million passengers can be derived, hence the GFA of 658,700m² for T2 (functional area and non-functional area).

3.3.2 Master Plan Works of the Terminal Area

1. Landside ground transportation system. The design should meet the landside transportation demands of 45 million passengers for T2 with relatively high service level. The construction size and standards should tally with the transportation demands and functional positioning of roads with optimal applicability and cost effectiveness. The construction work of landside transportation system of T2 mainly covers 28 roads, including the north approach tunnel, the east ring, the departure viaduct, the west ring and the south-extending south viaduct. Two surface parking lots are provided on the east and west sides of the north approach tunnel on the south of GTC, covering a total area of 53,089 m² offering 1,528 private parking spaces.

2. Landside bridges. The construction work of landside utility bridges includes one main bridge and five ramp bridges, specifically the departure viaduct, the private car ramp of GTC, the south-extending south ramp, the east ramp of the north approach road, the east 3rd ramp and the west 3rd ramp.

3. Landside telecom pipe network. The north approach tunnel serves as a Category II tunnel while the bus and taxi tunnels fall in Category IV. The equipment to be installed mainly includes power distribution system, electrical lighting system, building lightning-proof and grounding system, fire protection system and ELV system.

4. Airside apron and taxiway. The construction work of airside apron includes the construction of new remote and contact-stand aprons of this phase and the renovation of corresponding taxiway system. The design provides two taxiways for Code F aircrafts between East 5th Pier and East 6th Pier, two taxiways for Code C aircrafts between West 4th Stopover Stand and West 5th Pier, four taxiways for Type C aircrafts between West 5th Pier and West 6th Pier, one taxiway for Code C aircrafts between East 6th Pier and North Vertical Contact Taxiway and another for Code E aircrafts between West 6th Pier and North Vertical Contact Taxiway, and two taxiways for Code E aircrafts on the apron north of the main building.

3.3.3 Construction Works of T2

1. T2. Located on the north of T1, T2 is a super-large international hub terminal with designed annual passenger volume of 45 million. The current expansion project with 78 stands (64 contact stands) is designed as a 44.5 high, 658,700m² large four-storey building with long-span steel reinforced concrete structure.

2. GTC and parking building. Located on the south of the main building of T2, GTC and parking building totals a floor area of 208,400 m² in a height of about 12.95 m. It is designed in reinforced concrete frame structure with two floors underground and three floors aboveground to realize organic interchange between different landside transportation means and convenient parking for passengers.

3. Reserved Auto People Mover (APM) system. As per Class C service standards of IATA, APM system should be provided if the walking distance exceeds 750 m. As the walking distance between the center of T2 entrance and the farthest pier is 690 m, spaces and structural loads for APM system is reserved to provide high-level service for future departure, arrival and especially transfer passengers, and facilitate the development of a convenient integrated airport hub together with T1.

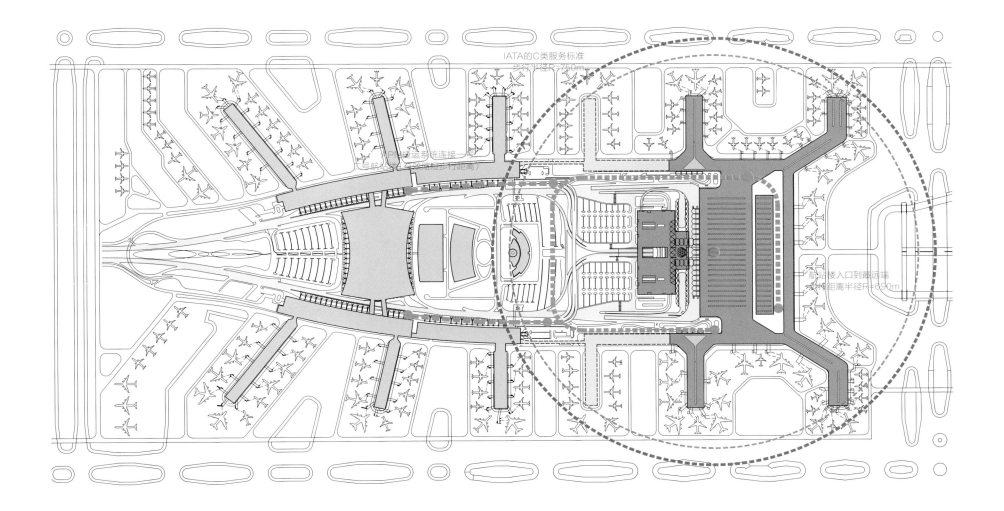

航站楼步行距离及旅客捷运系统预留
Walking Distance and Reserved APM System of T2

- - - 航站楼入口到最远端步行距离半径R=690m
Radius of Walking Distance from Entrance to the Farthest End of T2, R=690m

● 航站楼中间入口
Middle Entrance of T2

- - - IATA的C类服务标准，步行半径R=750m
Class C Service Standards of IATA, Walking Radius R=750m

● APM站台
APM Platform

- - - APM捷运系统连接一号、二号航站楼（有效缩短步行距离）
APM System Connecting T1 & T2 (for Shorter Walking Distance)

3.4 工程设计亮点

3.4 DESIGN HIGHLIGHTS

以打造世界级航空枢纽为目标，着力构建平安、绿色、智慧、人文的现代化体验式航站楼，实现可持续发展。

3.4.1 创新前置并行、灵活共享的安检大厅

通过对联检模式的研究，结合空间组合与功能流线，提出了以旅客流线为核心的联检布置体系，创新性地将国际、国内安检厅并置和前置，将国际、国内安检并排设置在第一关，利用国际、国内高峰小时错峰的特点，对安检通道国际国内功能进行切换，解决特殊高峰时段（如春节）流量瓶颈问题。

3.4.2 创新国内面积最大、近机位数量最多的"混流"流程与便捷枢纽机场中转流程

国内流程采用出发到达混流模式，相比传统出发、到达分流设计，既减少了独立的到达楼层通道和服务机房面积，又共用了商业、卫生间、问询柜台等服务设施，还减少了登机口管理人员，降低了投资，提高了资源利用率。同时混流设计让国内转国内这一中转流程在平层中转，不需换层，提高了中转效率。重点打造国际枢纽机场中转流程，其中国内转国内、国际转国内流程平层解决，国内转国际、国际转国际流程指引清晰、便捷高效。

The design aims to create an excellent aviation hub with safe, green, smart and humanistic terminals offering modern experience and enabling sustainable development.

3.4.1 Juxtaposed and Shared Security Check System at the First Pass

Based on researches on CIQ mode, spatial combination and functional circulation, the design introduces a passenger flow-centered CIQ layout system, placing the international and domestic security check halls in an innovative juxtaposed way at the first pass. During special peak-traffic period (like the Spring Festival), such design can realize conversion between international and domestic functions thus avoid congestion in the security check channel given that the peak traffic of international and domestic passengers usually occur in different hours.

3.4.2 The Largest Domestic Mixed-flow Area with the Most Contact Stands and Convenient Transfer Flows

The design adopts mixed domestic departure and arrival flows, which compared with the traditional approach of separate departure and arrival flows, downsizes the independent arrival floor channel and service machine room, and realizes sharing of such service facilities as commerce, washrooms and inquiry counters. In addition, it reduces the number of boarding management staff, cuts the investment and increases resource efficiency. Meanwhile, the mixed-flow design enables same-floor domestic-domestic transfer thus improves the transfer efficiency. With key consideration given to the transfer flows of the Project as an international hub airport, the design realizes domestic-domestic and international-domestic transfers on the same floor, and clear and efficient orientation for domestic-international and international-international transfers.

特色混流流程（西、北指廊）
Featured Mixed Flow (West & North Piers)

特色联检流程（安检大厅）
Featured CIQ Flow (Security Check Hall)

3.4.3 设置特色国际国内"可转换机位"与"登机桥固定端"（含A380）

迎合南航未来机队的机型组合系统及发展需要，提供最高效的机坪与机位配置，满足南航作为基地与枢纽航空公司运营需求。引入可在国内与国际运营之间进行转换的混合机位概念，将流线最便捷、流程最短的前列式黄金大机位设置成国内/国际可切换的混合机位，通过登机桥固定端实现机位与航站楼各进出港层的对接，提高机位使用灵活性及旅客处理效率，满足航司业务快速增长的需要。

3.4.3 Swing Stands and Fixed Ends for Boarding Bridges (incl. A380)

To accommodate the future aircraft mix and development demands of CSZ as a base and hub airline, the design gives the most efficient apron and stand configuration. It introduces the concept of swing stands, which allow for flexible conversion between domestic and international functions in the most valuable linear stand area that guarantees the most convenient and shortest flows. The boarding bridges are designed with fixed ends to realize connection between stands and various arrival/departure floors of the terminal. This can greatly improve flexible service of the stands and efficient passenger handling, thus keep up with the rapid growth of the aviation business.

特色"可转换机位"与"登机桥固定端"
Featured "Swing Stands" and "Fixed Ends for Boarding Bridges"

3.4.4 创新国内首个DCV小车自动分拣行李系统,并采用多种旅客智能设备

行李系统采用国际先进的DCV小车自动分拣技术,分拣度快、出错率低、效率高。值机行李交运部分采用自助模式,并为传统与自助系统切换做了预留设计。结合自助办票、自助托运、自助安检和自助登机等智能设备,实现无感化通关,为实现全面自助的智慧机场发展创造了条件。

3.4.5 创新特色文化机场

在位于航站楼内部安检、联检区北侧区域,设置可供旅客休憩的富有岭南特色的屋面内庭院——岭南花园。让旅客在现代化的航站楼内可以感受到传统园林的魅力,是人与自然融合的最佳实践,身处这里,犹如漫长旅途的一片绿洲,体现了广州地处中国南方的气候特点。北指廊的国际到达旅客可以欣赏到岭南花园的美景,留下对广州美好的第一印象。在办票大厅核心区域设置文化广场,为机场营造展示区域,为旅客提供观赏休憩空间。

特色行李系统(DCV小车自动分拣)
Featured Baggage System (DCV-Aided Auto Baggage Sorting)

特色岭南花园
Featured Lingnan Gardens

3.4.4 China's First DCV-aided Auto Baggage Sorting System and Smart Passenger Devices

The design introduces DCV-aided auto baggage sorting system, an internationally leading technology with fast sorting speed, low error rate and high efficiency. It adopts self-service baggage check-in and reserves room for switch between traditional and self-service systems. Smart self-service devices for check-in, baggage check, security check and boarding are also adopted to realize easy pass and possibly a smart airport with comprehensive self-service functions.

3.4.5 Featured Cultural Airport

Lingnan gardens, inner courtyards with Lingnan characteristics, are provided on the north of the security check and CIQ inspection areas inside the Terminal for passengers to take a rest. They allow passengers to experience the charm of traditional gardens in a modern terminal setting. This is the best practice to integrate man and nature. Being here is like encountering an oasis during a long journey, and passengers can experience the climatic features of Guangzhou as a city in southern China. International arrival passengers from the north pier can enjoy the beautiful scenery of Lingnan gardens and make a good first impression on Guangzhou. A cultural square is provided in the core area of the check-in hall for cultural display and rest.

可调节电动遮阳百叶内景
Interior View of Adjustable Electric Louvers

特色绿色机场（太阳能光伏发电）
Featured Green Airport (Solar PV Generation)

3.4.6 创新国内运行规模最大的绿色三星航站楼

充分考虑项目及所处区域气候特点，综合采用多种绿色建筑技术，力求达到资源、能源的最大化利用，创造高效、健康、节能、舒适的新型绿色机场建筑。采用可调节电动遮阳百叶、太阳能光伏发电、自然天窗采光通风、雨水收集利用、虹吸排水、变风量控制制冷、热回收空调机组、能源管理、建筑设备监控、智能照明控制等绿色建筑技术，荣获"国家三星级绿色建筑设计标识"。

3.4.6 The Largest 3-star Green Terminal in China

With full consideration of the Project characteristics and the climate of Guangzhou, the design adopts multiple green building technologies for T2 in a bid to maximally utilize resources and energy sources and create an efficient, healthy, energy-conserving and pleasant new green airport. Supported by such green building technologies as adjustable electric sun-shading louvers, solar PV generation, lighting and ventilation through skylights, rainwater harvesting and recycling, siphon drainage, AC units with VAV controlled cooling and heat recovery, energy management, building services monitoring and intelligent lighting control, the Project has obtained the Certificate of National 3-Star Green Building Design Label.

特色绿色机场（可调节电动遮阳百叶）
Featured Green Airport (Adjustable Electric Louvers)

特色集中商业（国际商业区）
Featured Centralized Commerce (International Commercial Area)

3.4.7 创新结合流程设置的航站楼非典型商业

根据国际、国内机场的不同案例和对商业综合体的研究，提出了独特的航站楼商业概念——非典型商业综合体。典型的综合体从规划层面就确定了出入口位置及数量、人流动线模式及、业态布局分类。而航站楼的重中之重是旅客流程概念规划基于各种流程开始，功能也服务于流程。相对典型商业综合体，航站楼具有旅客聚散集中、旅客流动性高、停留时间短的不同特点。因此，为了达到增加非航收入的目标，在和商业顾问共同研究之后，从整合旅客流程、活跃商业动线、协调专业资源、合理增加面积，创造跃层中庭、营造商业氛围、组合功能布局、控制业态配比，引入生态绿化、优化商业环境等五个方面创造了不同于国内其他枢纽机场的航站楼商业布局模式。

3.4.7 Non-typical Flow-oriented Commercial Layout

Based on cases of international and domestic airports and researches on commercial complex, the design proposes a unique concept of non-typical commercial complex. For a typical complex, its entrance/exit location and quantity, pedestrian circulation mode and trade layout are determined right in the planning stage. However, for a terminal whose top priority is passenger flows, its conceptual planning and functional configuration are both based on various flows. Different from a typical commercial complex, a terminal building features centralized passengers with high mobility and short stay. Therefore, to increase non-aviation revenue, after consulting the commercial consultant, the design creates a unique commercial layout for T2 that is different from other domestic hub airports through five measures. These measures include integrating passenger flows and providing dynamic commercial circulations; coordinating disciplinary resources to increase the floor area; creating a skip-floor atrium to foster strong commercial atmosphere; combining functions and controlling trade mix; and introducing ecological greening for better commercial environment.

特色集中商业（国内商业区）
Featured Centralized Commerce (Domestic Commercial Area)

第4章

构型设计与交通组织
CHAPTER 4 TERMINAL LAYOUT AND TRAFFIC ORGANIZATION

4.1 方案推演
4.2 空侧规划保证飞机运行效率
4.3 与空侧容量相匹配的陆侧交通设计

4.1 EVOLUTION OF SCHEMES
4.2 AIRSIDE PLANNING FOR EFFICIENT AIRCRAFT OPERATION
4.3 LANDSIDE TRAFFIC COMPATIBLE WITH AIRSIDE CAPACITY

4.1 方案推演

4.1.1 航站区总体规划方案演变过程

广州白云国际机场总平面构型构思于20世纪90年代末期,1999年11月民航总局以民航机函[0999]798号文《关于广州白云国际机场总平面规划的批复》批准了项目的总平面规划。原总体规划年旅客吞吐量为5200万人次。

为适应广州白云国际机场的定位及航空市场业务发展的需求调整,广州白云国际机场总平面构型同步在不断地优化,前后历经了分离式站坪方案、中滑方案及北站坪方案等三个主要阶段。

1. 分离式站坪方案。二号航站楼位于一号航站楼北侧,一、二号航站楼在南北两侧都设置有陆侧道路系统,整个航站区形成明确分隔的东西站坪,东西站坪依靠南北两端的飞机滑行联络道联系。一号航站楼自2004年8月通航运营以来,东西站坪联系距离过远导致空侧运行效率偏低的问题凸显。因此采取分离式站坪方案会进一步加长东西站坪的联系距离,从而进一步降低空侧运行效率。

2. 中滑方案。在二号航站楼南侧3指廊和4指廊之间增加两条东西向联络道,强化东西站坪间的联系,使得东西站坪间航班的调度更为灵活,滑行距离更为简捷。新建的垂直联络滑行道使得更多的机位位于东、西主跑道中间的位置,临近区域机位(航站区中央范围)双向运行便捷程度大大提升,有利于提高飞行区与航站区之间的运营效率。

3. 北站坪方案。在二号航站楼北侧设置北站坪,通过北站坪将东、西站坪连为一整体,同时北站坪北侧布置两条机坪滑行通道,这种布局方式有利于站坪机位的灵活运行及提高飞机在东、西站坪上的运行效率。另外在北站坪方案中,二号航站楼的近机位和远机位中D类及以上的大型机位所占的比例为50%,对未来航空市场机型变化有更大的灵活性。

4.1.2 方案优化比选结论

分离式站坪方案、中滑方案及北站坪方案每个方案都有其各自独特的优势和劣势。经过多轮的方案评估,以及重点结合南航枢纽运作需求的仔细研究分析,最后基于下列主要优点决定选用北站坪方案作为最终二号航站楼实施方案:(1) 近机位数量最多;(2) 国际及国内登机门以及所有指廊之间(包括一号航站楼指廊)的方便连接;(3) 东、西站坪可以通过北站坪高效连接;(4) 简化的道路系统及航站楼流程;(5) 通过规划的APM系统实现灵活的未来扩展性;(6) 最大限度保留原有的基础设施。

4.1 EVOLUTION OF SCHEMES

4.1.1 Evolution of the Master Plan of Terminal Area

The master plan of Guangzhou International Airport (Baiyun Airport) was conceived in the late 1990s and approved by the Civil Aviation Administration of China in the *Approval of the Master Plan of Guangzhou Baiyun International Airport (Min Hang Ji Han [0999] No. 798)* in November 1999. The original planned annual passenger traffic was 52 million.

To adapt to the positioning of Baiyun Airport and the development demands of the aviation business, the master plan has been continuously optimized through three main stages: separated apron scheme, middle taxiway scheme, and north apron scheme.

1. Separated apron scheme. This scheme places Terminal 2 (T2) on the north of Terminal 1 (T1), with landside road systems established on the south and north of both terminals. This creates clearly divided east and west aprons throughout the terminal area connected by the contact taxiways provided in the south and north ends. Since T1 became operational in August 2004, low airside operation efficiency caused by the long distance between the east and west aprons has become increasingly prominent. Therefore, the separated apron scheme would only further extend the distance between the two aprons, resulting in lower efficiency of airside operation.

2. Middle taxiway scheme. This scheme adds two east-west contact taxiways between Pier 3 and Pier 4 on the south of T2 to strengthen the connection between the east and west aprons hence more flexible schedule of flights and shorter taxiing distance. The new vertical contact taxiways concentrate more stands in the middle of the east and west main runways. This would enable convenient two-way communication between stands nearby (within the central area of terminal area), and significantly improve the efficiency of operation between the movement area and the terminal area.

3. North apron scheme. This scheme proposes a north apron on the north of T2 to integrate the east and west aprons, and arranges two taxiways in the north of the north apron. This layout is conductive to the flexible operation of stands and can improve the operation efficiency of aircrafts on the east and west aprons. In addition, this scheme allocates 50% of the contact and remote stands of T2 to Category D and upper-level aircrafts, which makes it more flexible to adapt to the future model changes in the aviation market.

4.1.2 Conclusion of Scheme Comparisons

Each of the above three schemes has its own advantages and disadvantages. After several rounds of scheme evaluation and in consideration of the hub operational demands of China Southern Airlines (CSN), the north apron scheme was selected for T2 due to the following merits: (1) It offers the most contact stands; (2) It enables convenient access between the international and domestic boarding gates and all piers (including those of T1); (3) The east and west aprons are efficiently connected by the north apron; (4) It streamlines the road systems and flows in the terminal; (5) It ensures flexible scalability in the future through the planned APM system; (6) It retains the original infrastructure to the maximum extent.

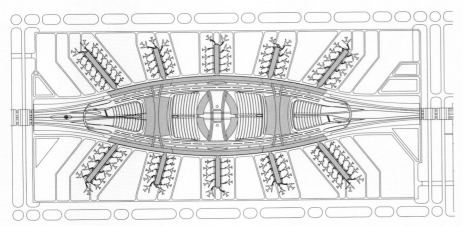

航站楼规划方案（1999年）
Terminal Planning Scheme (1999)

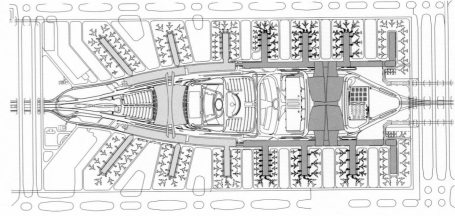

分离式站坪方案
Separated Apron Scheme

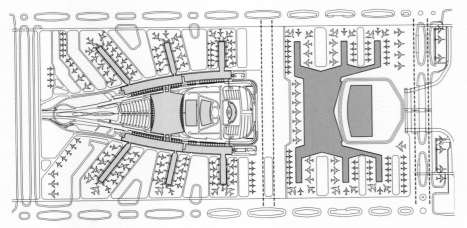

中滑方案
Middle Taxiway Scheme

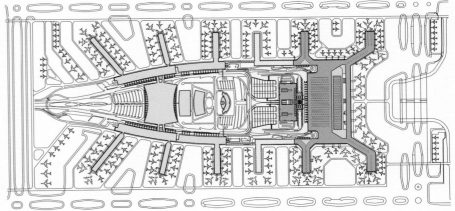

北站坪方案
North Apron Scheme

4.2 空侧规划保证高效的飞机运行
4.2 AIRSIDE PLANNING FOR EFFICIENT AIRCRAFT OPERATION

1. 空侧规划原则。适应航站楼的设计整体,充分利用现有土地资源;提供最高效的机坪与机位配置,满足南航作为基地与枢纽航空公司运营需求。
2. 混合机位概念。将流线最便捷、流程最短的前列式机位设置成国内/国际可转换的混合机位,提高机位使用灵活性及旅客处理效率,满足国际业务快速增长的需要。
3. 组合机位概念。为实现机位布置最大化与提升空侧机坪使用灵活度,设计引入可供大飞机与小飞机同时停靠的组合机位,即可供一架E类、F类飞机或两架C类停靠的机位,达到机位满足各类飞机错时靠港的使用需求。
4. 保证近机位数量。二号航站楼一期工程包括64个近机位,其中国内机位共34个(31C、3E),国际机位共21个(8C、11E、2F),国内与国际可转换的混合机位共9个(7E、2F);二期建设中扩建将再增加26个近机位,总体规划共90个近机位。
5. 双滑行道的指廊间距。指廊与指廊之间按照同类机型布置机位,用双滑行道保证滑行效率,并节省空间。

1. Airside planning principles. Adapt to the holistic terminal design to fully utilize existing land resources, and reference the future aircraft mix provided by CSN to meet its development needs; provide the most efficient apron and stand configuration to meet CSN's operation needs as a base and hub airline.
2. Swing stands. Linear stands with the most convenient and shortest flows are designed into swing stands to enhance flexible stand operation and efficient passenger handling and meet the demands of rapidly growing international airline business.
3. MARS stands. To maximize the number of stands and the flexibility of airside apron operation, MARS stands are introduced for the parking of both large and small aircrafts. Each MARS stand can accommodate a Code E and a Code F aircraft or two Code C aircrafts at the same time, allowing various types of aircrafts to park during staggered time periods.
4. Sufficient contact stands. Phase I of T2 offers 64 contact stands, including 34 domestic ones (31C, 3E), 21 international ones (8C, 11E, 2F) and 9 swing ones (7E, 2F). Phase II will include another 26 contact stands, making a total of 90 contact stands.
5. Pier interval for double taxiways. Stands for the same aircraft type are provided between piers, with double taxiways to ensure the efficiency of taxiing and space.

C类飞机 Code C Aircraft
E类飞机 Code E Aircraft
F类飞机 Code F Aircraft

空侧机位组合图
Airside Stand Combo

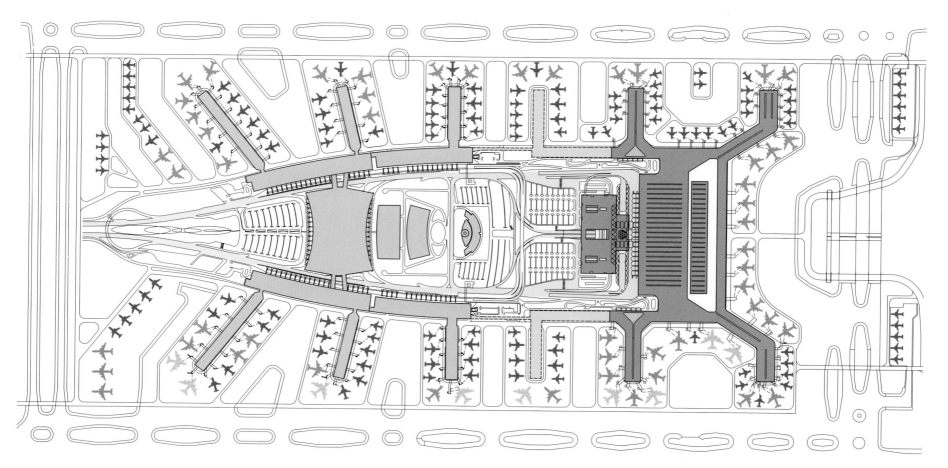

4.3 与空侧容量相匹配的陆侧交通设计

陆侧交通一体化,实现地铁、城轨、大巴、出租车等各种交通方式无缝连接,平层解决主要交通换乘,缩短换乘步行距离。

1. 南北贯通道路交通。陆侧设置贯通南北的道路系统,确保白云机场与外部城市交通衔接的安全性。通过东西环路实现一号、二号航站楼分别与南北出入口的快捷联系及两楼的交通互不干扰,减少车辆长距离绕行给机场环路带来的交通压力,达到了既有效分流车辆,又保障道路畅通的目的。

2. 交通中心及停车楼。以"公共交通优先、大众旅客优先、离港旅客优先"为设计原则,公共交通(地铁、城轨、市区大巴、长途大巴、出租车)前置,缩短了大部分旅客步行距离。地铁站厅和城轨站厅分列交通中心大厅的东西两侧,有效分流旅客。私家车停车楼紧邻交通中心,旅客可通过垂直交通核心到达各层车库。

3. 与轨道交通衔接。白云机场配套了两条地下轨道即地铁三号线和城际轨道线,均在交通中心地下二层设置站厅。地铁、城轨的出发旅客和到达旅客与航站楼均有各自的流线设计,互不干扰;两轨道交通之间设有专用的换乘通道,高效便捷。

4. 人车分流的陆侧交通。二号航站楼到达层(首层)迎客厅与交通中心之间设置开阔的行人广场;航站楼前的到达大巴车道及出租车道均采用下穿隧道的方式,实现完全的人车分流,旅客可以方便地平层步行到达市区大巴上客区、长途大巴候车区、私家车停车楼,或通过垂直交通到达地铁站厅、城轨站厅及停车楼其他楼层。

基于《广州白云国际机场二号航站楼及配套设施规划设计咨询报告》对于航班客流量的预测,对目标年(2020年)高峰小时客流量、高峰小时陆侧交通流量及高峰小时陆侧交通停车需求进行预测,具体详见本章节表格。

1. 交通中心停车楼 1. GTC Parking Building
2. 停车场 2. Parking Lot
3. 巴士站场 3. Bus Station
4. 南北进出场车流 4. South and North Access Vehicular Flows
5. 陆侧高架桥 5. Landside Viaduct

陆侧交通流线
Landside Traffic Circulation

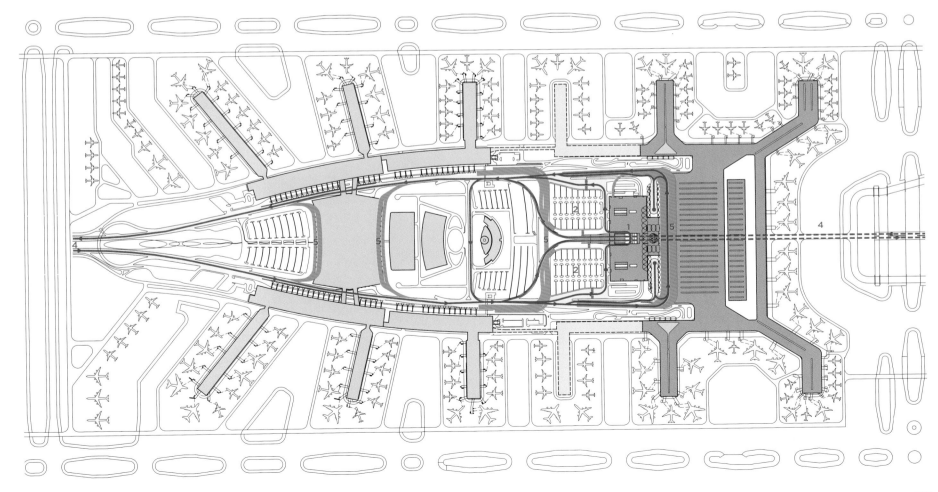

4.3 LANDSIDE TRAFFIC COMPATIBLE WITH AIRSIDE CAPACITY

Integrated landside transportation is provided to realize seamless interchange between all means of transportation including metro, intercity railway, bus, taxi etc., with major interchanges organized on the same level to reduce walking distance.

1. North-south road traffic. A north-south road system is provided on landside for safe traffic connection between Baiyun Airport and external urban transportation. Convenient connection between the north and south accesses and T1 and T2 are established respectively via east and west ring roads to avoid mutual interference and reduce the pressure of long detour on airport ring roads. This way the vehicular traffic can be effectively diverted and smooth road conditions can be guaranteed as well.

2. GTC and parking building. GTC offers passengers convenient interchange between various means of transportation. It is designed based on the principle of "prioritizing public transportation, common passengers and departure passengers". The public transportation facilities (metro, intercity railway, city bus, long distance coach, taxi) are most conveniently located to reduce the walking distance of most passengers. Concourses of metro station and intercity railway station are provided respectively in the east and west sides of GTC concourse, effectively diverting passenger flows. Parking building for private cars is right next to GTC, so passengers can reach all garage floors via a vertical transportation core.

3. Connection to rail transit. Baiyun Airport is connected with two underground rail lines, i.e., Metro Line 3 and an intercity rail line, both stationed on B2 in GTC. From metro and intercity railway to terminal, separate circulations are designed for departure and arrival passengers to avoid mutual interference. Besides, dedicated access is designed between the two lines of rail transit for great efficiency and convenience of interchange.

4. Landside transportation with separated pedestrian and vehicular circulations. A spacious pedestrian square is provided between the arrival hall on arrival floor (F1) in T2 and GTC, while the arrival bus and taxi lanes in front of the terminal are provided in the form of underground tunnels, realizing complete separation of pedestrian and vehicular circulations. Therefore, passengers can conveniently access city bus pick-up area, long-distance coach waiting area, and private car parking building on the same level or access metro station concourse, intercity railway station concourse and other floors in parking building etc. via vertical transportation.

Based on the forecast about passenger traffic specified in the *Consultation Report on the Planning and Design for T2 and Supporting Facilities in Guangzhou Baiyun International Airport*, forecasts about the passenger traffic, landside traffic and landside parking during peak hours in the target year (2020) should be provided. For details, please refer to the table in this chapter.

1. 广州白云国际机场　　1. Guangzhou Baiyun International Airport
2. 南进场路　　　　　　2. South Approach Road
3. 北进场路　　　　　　3. North Approach Road
4. 机场高速环线　　　　4. Airport High-Speed Ring
5. 机场高速　　　　　　5. Airport Expressway
6. 第二机场高速　　　　6. 2nd Airport Expressway

车辆分类 Vehicle Classification	旅客乘车比例 Passenger Proportion	每车载客数 Capacity Per Vehicle	满载率 Load Factor	高峰小时旅客量（人次/小时） Peak-Hour Passenger Traffic (Person / Hour)		
				全机场 Total	一号航站楼 T1	二号航站楼 T2
轻轨 Light Rail	25%			4776	2514	2262
大巴 Bus	20%	40	0.8	3821	2011	1810
中巴 Mini-Bus	5%	8	1	955	503	452
私家车 Private Car	25%	1.5	1	4776	2514	2262
出租车 Taxi	20%	1.5	1	3821	2011	1810
停车场班车 Shuttle Bus	3%	20	1	573	302	271
其他 Other	2%	40	0.8	382	201	181
合计 Total	100%			19105	10055	9050

2020年机场高峰小时客流量
Peak-hour Passenger Traffic in 2020

车辆分类 Vehicle Classification	全机场 Total		一号航站楼 T1		二号航站楼 T2	
	接人 Pick up	送人 See off	接人 Pick up	送人 See off	接人 Pick up	送人 See off
轻轨 Light Rail	–	–	–	–	–	–
大巴 Bus	60	60	31	31	28	28
中巴 Mini-Bus	60	60	31	31	28	28
私家车 Private Car	1592	1592	838	838	754	754
出租车 Taxi	1274	1274	670	670	603	603
停车场班车 Shuttle Bus	14	14	8	8	7	7
其他 Other	6	6	3	3	3	3
合计 Total	3005	3005	1582	1582	1424	1424
机场员工车辆 Vehicles of Airport Staff	150	150	79	79	71	71
机场服务设施车辆 Vehicles for Airport Service	60	60	32	32	28	28
总计 Total	3216	3216	1693	1693	1523	1523

2020年机场高峰小时陆侧交通量
Landside Peak-hour Passenger Traffic in 2020

车辆分类 Vehicle Classification	全机场 Total		一号航站楼 T1		二号航站楼 T2	
	接人 Pick up	送人 See off	接人 Pick up	送人 See off	接人 Pick up	送人 See off
预测高峰小时交通量（标准车/小时） Forecast Peak Hour Traffic (Standard Car / Hour)	3285	3285	1729	1729	1556	1556

2020年机场高峰小时陆侧交通量
Landside Peak-hour Passenger Traffic in 2020

车辆分类 Vehicle Classification	高峰小时车位需求 Peak-Hour Parking Space Demand		
	一号航站楼 T1	二号航站楼 T2	全机场 Total
大巴 Bus	102	131	233
中巴 Mini-Bus	102	131	233
私家车 Private Car	3653	4697	8350
短期 Short Term	1372	1764	3136
长期 Long Term	2281	2933	5214
出租车 Taxi	1199	1542	2741
工作人员 Staff	275	353	628
总计 Total	5331	6854	12185

2020年机场高峰小时停车需求预测
Peak-hour Parking Demand Projection in 2020

第5章
流程设计与功能配置
CHAPTER 5 TERMINAL FLOW AND FUNCTIONAL CONFIGURATION

5.1　概述
5.2　功能分布与流程设计
5.3　功能配置

5.1　GENERAL
5.2　FUNCTIONAL DISTRIBUTION AND FLOW DESIGN
5.3　FUNCTIONAL CONFIGURATION

5.1 概述

5.1.1 二号航站楼较一号航站楼流程优化及容量对比

一号航站楼的特点是出发和到达分开。办票大厅集中办票,然后再分散A、B区登机,到达同样分A、B区,国际航班及非南航国内航班在A区(东侧),南航的国内航班在B区(西侧)。其优点是旅客分流,不会出现过分集中的旅客流。其外部的交通系统也是一种比较新的理念,整个航空规划的特点是多指廊式构型,可以停靠较多的飞机。

二号航站楼不再采用"A、B"区,而是采用单一方向策略。构型根据飞机空侧运行效率的需要进行调整,由原来类似一号航站楼的分离站坪概念调整为北站坪概念,把机场北面的进场路下沉穿过机坪及航站楼与南面路网连接,使得机坪在东、西及北三面连通,便于飞机调度,增加近机位,特别是与主楼路程最短的大型近机位,缩短了旅客平均步行距离。二号航站楼的功能配置更为全面,涵盖旅客各个层面的需求,对于计时休息、儿童活动等特殊功能与空间需求都做了综合考虑。

白云机场设计年旅客吞吐量为2020年8000万人次,其中一号航站楼为3500万人次,二号航站楼为4500万人次。

5.1.2 设计原则

二号航站楼是大型国际枢纽机场航站楼,拥有众多不同类型的复杂流程,作为交通枢纽建筑,简洁和高效是流程设计的核心。根据业主确定的二号航站楼主要供以基地航空公司南航为主的天合联盟使用的定位,结合枢纽机场的要求,确定了二号航站楼流程设计的原则:

1. 以旅客体验为导向,流程简洁、高效,最大限度地满足旅客需求,提升旅客出行体验;
2. 覆盖并满足基地航空公司南航及天合联盟的所有航空产品需求;
3. 最大限度满足现代枢纽机场航站楼运营保障需求。

5.1.3 功能与流程构成

作为一个大型的复合枢纽机场航站楼,二号航站楼拥有各种类型的旅客流程,包括国内出港、国内进港、国际出港、国际进港、国内转国际、国内转国内、国际转国内、国际转国际、国际航班国内段(国内-广州-国际、国际-广州-国内、广州-国内-国际、国际-国内-广州)等,各流程或交叉或联系或结合又形成新流程,错综复杂。设计采取了国内混流、国际分流、设置混合机位的策略来搭建航站楼的内部组织框架和楼层。

5.1 GENERAL

5.1.1 Flow & Capacity Comparisons between T2 and T1

Terminal 1 (T1) is characterized by its separation of departure and arrival flows. Passengers first gather in the check-in hall for check-in and then head for Zone A or B for boarding. The arrival hall is also divided into Zone A and B: passengers of international flights and non-China Southern Airlines (CSZ) flights arrive at Zone A (in the east), while passengers of domestic flights of CSZ at Zone B (in the west). Such design can distribute passenger flows and avoid over-concentration. The external transportation system of T1 has adopted a relatively new concept, i.e. multi-pier layout that allows more aircrafts to park in.

The layout of Terminal 2 (T2) employs a one-way strategy without providing Zone A and B. For higher efficiency of airside aircraft operation, the layout was adjusted from separated apron similar to T1 to north apron. The new design sinks the approach road on the north of the Airport to underneath the apron and the terminal to connect it with the road network on the south. That connects through the apron on the east, west and north sides, making it convenient for flight scheduling and creating more contact stands, especially large ones closest to the main building, to shorten the average walking distance for passengers. Functional facilities of T2 are more inclusive as they cover all aspects of passengers' demands and have taken comprehensive consideration of the special functions and spatial requirements such as hourly charged lounge and children activities zones.

The designed annual passenger traffic of Baiyun Airport is 80 million by 2020, with 35 million for T1 and 45 million for T2.

5.1.2 Design Principles

T2 is a defined as a terminal of a large international hub airport with different types of complicated flows. For such a transportation hub building, simplicity and efficiency is the core of its flow design. As T2 is positioned to serve mainly the SkyTeam members, especially the based airline CSZ, its flow design principles are determined a below in consideration of the hub airport requirements:

1. Be passenger-oriented, simple and efficient, satisfy the demands of passengers at maximum level to improve their travel experience;
2. Cover and satisfy the requirements of all airline products of the base airline CSZ and the Sky Team;
3. Satisfy the operational support demands of a modern hub airport terminal at maximum level.

5.1.3 Function and Flow Composition

As a large compound hub airport terminal, T2 accommodates all types of passenger flows, including domestic departure, domestic arrival, international departure, international arrival, domestic-international transfer, domestic-domestic transfer, international-domestic transfer, international-international transfer, and domestic segments of international flights (domestic-Guangzhou-international, international-Guangzhou-domestic, Guangzhou-domestic-international, international-domestic-Guangzhou) etc. Some flows intersect with each other, some interconnected, and some combined into new flows. The design adopts strategies of domestic mixed flows, international separate flows and swing stands to construct the internal frame and floors of the terminal.

5.2 功能分布与流程设计 / 5.2 FUNCTIONAL DISTRIBUTION AND FLOW DESIGN

5.2.1 功能分布 / 5.2.1 Functional Distribution

楼层分布 Floor	主要功能 Main Functions
负一层平面 B1	主楼：进港行李下送地沟、水电空设备专业管沟 指廊：水电空设备专业管沟 Main building: Below-grade delivery trench for arrival baggage, ditches for plumbing, electrical and AC pipelines Pier: Ditches for plumbing, electrical and AC pipelines
一层平面 F1	主楼：国内国际迎客厅、国内国际行李提取厅、国际国内中转提取行李交运厅、国内国际行李分拣机房、海关/检验检疫查验通道及办公用房、内部办公用房、配套商业/服务设施用房、相关水电空设备专业用房 指廊：国内及国际贵宾室、国内及国际远机位候机厅、国内及国际误机旅客等候厅、国际航班国内段的进港到达厅/出港候机厅、海关/检验检疫空侧办公用房、空侧服务用房、空侧机坪服务用房、行李机房配套用房、其他配套内部办公用房、设备用房 Main building: Domestic and international arrival hall, domestic and international baggage claim hall, international-domestic transfer baggage recheck-in hall, domestic and international baggage sorting machine room, customs/CIQ check channel and office rooms, internal office rooms, supporting commercial/service facilities, relevant plumbing, electrical and AC rooms Pier: Domestic and international VIP rooms, domestic and international remote stand waiting hall, domestic and international no-show waiting hall, arrival hall / departure waiting hall for domestic segment of international flights, airside customs / CIQ office rooms, airside service room, airside apron service room, supporting baggage machine room, other supporting internal office rooms, MEP rooms
二层平面 F2	主楼：国际进港旅客卫检厅、入境边防大厅、国际转国际中转厅及安检厅、国际转国内（通程联运）中转厅及安检厅、入境免税商业区、国内旅客进港通道、国内混流出发及到达集中商业区、内部办公用房、旅客服务设施、相关水电空设备专业用房 指廊：国内及国际进港旅客到达通道、国内出港旅客候机厅、卫生间、内部办公用房、设备用房 Main building: International arrival passenger inspection and quarantine hall, immigration border defense hall, international-international transfer hall and security check hall, international-domestic (through-check-in) transfer hall and security check hall, entry duty-free commercial area, domestic passenger arrival channel, centralized commercial area for mixed-flow domestic departure and arrival, internal office facilities, passenger service facilities, relevant plumbing, electrical and AC rooms Pier: Domestic and international arrival passenger arrival channel, domestic departure passenger waiting hall, washrooms, internal office rooms, MEP rooms
三层平面 F3	主楼：值机大厅（文化广场、值机岛）、安检大厅、联检大厅、相关配套服务设施（问讯、补办票、自助值机/托运、安检开包间、超大行李托运、客带货申报托运、两舱值机、商业/餐饮、航空公司办公用房、内部办公用房、两舱休息室、海关/边防/检验检疫办公用房、东翼国际大型国际免税商业区、西翼国内集中商业、餐饮及两舱休息室等） 指廊：国际出港旅客候机厅、国际进港旅客到达通道、配套商业服务设施、卫生间、内部办公用房、设备用房 Main building: Check-in hall (cultural square, check-in island), security check hall, CIQ hall, relevant supporting facilities (inquiry, ticket, self-service check-in/baggage check-in, baggage inspection room, extra-large baggage check-in, declared goods check-in, first class/business class check-in, commerce/F&B, airline office rooms, internal office rooms, first class/business class lounge, customs/immigration/CIQ office rooms, large international duty-free commercial area in the east wing, domestic centralized commerce, F&B and first class/business class lounge in the west wing) Pier: International departure passenger waiting hall, international arrival passenger arrival channel, supporting commercial service facilities, washrooms, internal office rooms, and MEP rooms
四层平面 F4	主楼：陆侧值机大厅集中餐饮、空侧国际两舱休息室、国际转国际计时休息室、航空公司及联检部门办公用房等 指廊：国内进出港旅客混流候机厅/到达通道、国际出港旅客候机厅、配套商业服务设施、卫生间、内部办公用房、设备用房 Main building: Centralized F&B in landside check-in hall, airside international first class/business class lounge, hourly charged lounge for international-international transfer passengers, airline and CIQ office rooms, etc. Pier: Mixed-flow domestic arrival and departure passenger waiting hall/arrival channel, international departure passenger waiting hall, supporting commercial service facilities, washrooms, internal office rooms, MEP rooms

二号航站楼平面功能分布表
Planar Function Distribution of T2

楼层分布 Floor	主要功能 Main functions
负二层平面 B2	私家车停车库（人民防空地下室）、设备中心、地铁站厅（他项工程，11.000m层）、城轨站厅（他项工程，16.000m层） Garage for private cars (civil air defense basement), MEP center, metro station concourse (other project, at Level -11.000m), intercity railway station concourse (other project, at Level -16.000m)
负一层平面 B1	私家车停车库（人民防空地下室） Garage for private cars (civil air defense basement)
一层平面 F1	私家车停车库、交通中心（旅客到达大厅、地铁3号线机场北站A出口，城轨二号楼站出口（未启用）、客运站（长途大巴、市区大巴候车区），中转大巴、航延大巴、旅游大巴等候区 Garage for private cars, GTC (passenger arrival hall, Exit A of Airport North Station of Metro Line 3), intercity railway T2 exit (not available), passenger station (waiting area for coaches and downtown buses), waiting area for transfer buses, buses for flight delay, tour buses
二层平面 F2	私家车停车库、交通中心（旅客通道、办公区、设备房） Garage for private cars, GTC (passenger channel, office area, MEP room)
三层平面 F3	私家车停车库、停车场办公区、停车场管理控制中心（TIC）、机房 Garage for private cars, parking lot office area, parking lot traffic information center (TIC), machine room
四层平面 F4	屋顶绿化私家车停车场、旅客通道 Garage for private cars on green roof, passenger channel

交通中心及停车楼平面功能分布表
Planar Function Distribution of GTC and Parking Building

5.2.2 流程设计

5.2.2 Flow Design

主要流程关系图
Relations between Main Flows

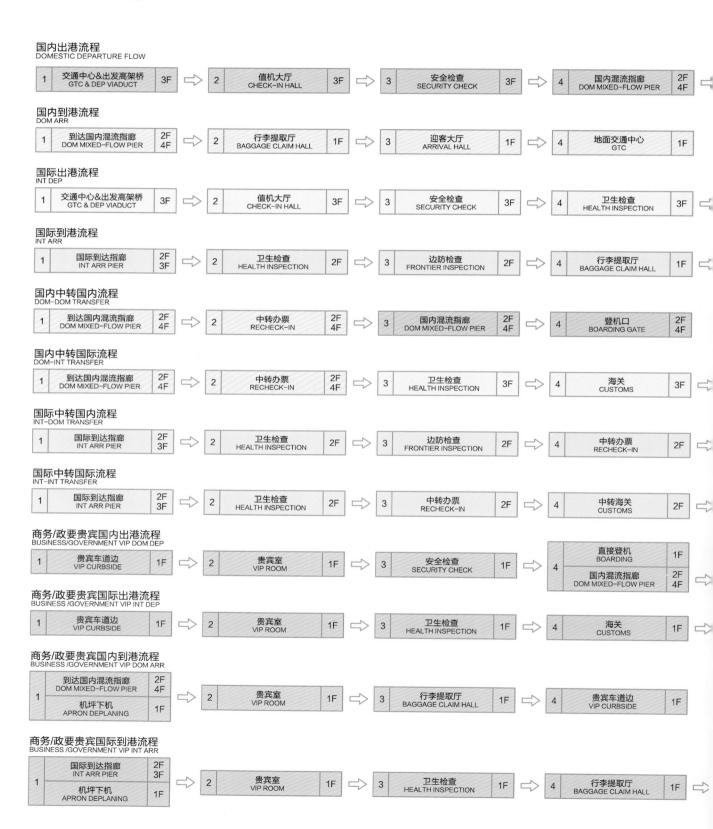

| 5 | 登机口 BOARDING GATE | 2F 4F |

| 5 | 海关 CUSTOMS | 3F | ⇒ | 6 | 边防检查 FRONTIER INSPECTION | 3F | ⇒ | 7 | 国际出发指廊 INT DEP PIER | 3F 4F | ⇒ | 8 | 登机口 BOARDING GATE | 3F 4F |

| 5 | 海关 CUSTOMS | 1F | ⇒ | 6 | 动植物检疫 ANIMAL & PLANT QUARANTINE | 1F | ⇒ | 7 | 迎客大厅 ARRIVAL HALL | 1F | ⇒ | 8 | 地面交通中心 GTC | 1F |

| 5 | 边防检查 FRONTIER INSPECTION | 3F | ⇒ | 6 | 国际出发指廊 INT DEP PIER | 3F 4F | ⇒ | 7 | 登机口 BOARDING GATE | 3F 4F |

| 5 | 中转海关 CUSTOMS | 2F | ⇒ | 6 | 中转安全检查 SECURITY CHECK | 2F | ⇒ | 7 | 国内混流指廊 DOM MIXED-FLOW PIER | 2F 4F | ⇒ | 8 | 登机口 BOARDING GATE | 2F 4F |

| 5 | 中转安全检查 SECURITY CHECK | 2F | ⇒ | 6 | 国际出发指廊 INT DEP PIER | 3F 4F | ⇒ | 7 | 登机口 BOARDING GATE | 3F 4F |

| 5 | 登机口 BOARDING GATE | 2F 4F |

| 5 | 边防检查 FRONTIER INSPECTION | 1F | ⇒ | 6 | 安全检查 SECURITY CHECK | 1F | ⇒ | 7 | 直接登机 BOARDING | 1F | ⇒ | 8 | 登机口 BOARDING GATE | 3F 4F |
| | | | | | 安全检查 SECURITY CHECK | 3F | ⇒ | 7 | 国际出发指廊 INT DEP PIER | 3F 4F | | | | |

| 5 | 海关 CUSTOMS | 1F | ⇒ | 6 | 边防检查 FRONTIER INSPECTION | 1F | ⇒ | 7 | 动植物检疫 ANIMAL & PLANT QUARANTINE | 1F | ⇒ | 8 | 贵宾车道边 VIP CURBSIDE | 1F |

其他流程关系图
Other Flow Relation Diagram

国际航班国内段流程（广州-国内-国际）
DOMESTIC SEGMENT OF INTERNATIONAL FLIGHT (GUANGZHOU-DOMESTIC-INTERNATIONAL)

| 1 | 交通中心&出发高架桥 GTC & DEP VIADUCT | 3F | ⇒ | 2 | 值机大厅 CHECK-IN HALL | 3F | ⇒ | 3 | 安全检查 SECURITY CHECK | 3F | ⇒ | 4 | 卫生检查 HEALTH INSPECTION | 3F | ⇒ |

国际航班国内段流程（国际-国内-广州）
DOM SEGMENT OF INT FLIGHT (INT-DOM-GUANGZHOU)

| 1 | 国际远机位到达走廊 INT REMOTE STAND ARR PIER | 1F | ⇒ | 2 | 行李提取厅 BAGGAGE CLAIM HALL | 1F | ⇒ | 3 | 海关 CUSTOMS | 1F | ⇒ | 4 | 动植物检疫 ANIMAL & PLANT QUARANTINE | 1F | ⇒ |

国际航班国内段流程（国际-广州-国内）
DOM SEGMENT OF INT FLIGHT (INT-GUANGZHOU-DOM)

| 1 | 国际到达指廊 INT ARR PIER | 2F 3F | ⇒ | 2 | 卫生检查 HEALTH INSPECTION | 2F | ⇒ | 3 | 边防检查 FRONTIER INSPECTION | 2F | ⇒ | 4 | 中转办票 TRANSFER RECHECK-IN | 2F | ⇒ |

国际航班国内段流程（国内-广州-国际）
DOM-INT TRANSFER (INT.PART CANTON DEP)

| 1 | 首层专用到达厅 DEDICATED ARR HALL | 1F | ⇒ | 2 | 卫生检查 HEALTH INSPECTION | 3F | ⇒ | 3 | 海关 CUSTOMS | 3F | ⇒ | 4 | 边防检查 FRONTIER INSPECTION | 3F | ⇒ |

国内取消航班旅客流程
DOM CANCELLED FLIGHT

| 1 | 国内混流指廊 DOM MIXED-FLOW PIER | 2F 4F | ⇒ | 2 | 行李提取厅 BAGGAGE CLAIM HALL | 1F | ⇒ | 3 | 迎客大厅 ARRIVAL HALL | 1F | ⇒ | 4 | 地面交通中心 GTC | 1F |

国际取消航班旅客流程
INT CANCELLED FLIGHT

| 1 | 国际出发指廊 INT DEP PIER | 3F 4F | ⇒ | 2 | 边防检查 FRONTIER INSPECTION | 3F | ⇒ | 3 | 海关 CUSTOMS | 3F | ⇒ | 4 | 卫生检查 HEALTH INSPECTION | 3F | ⇒ |

到港旅客转地铁
INTERCHANGE WITH METRO FOR ARRIVAL PASSENGERS

| 1 | 迎客大厅 ARRIVAL HALL | 1F | ⇒ | 2 | 交通中心旅客大厅 GTC PASSENGER HALL | 1F | ⇒ | 3 | 扶梯下至地铁站厅 METRO STATION | B1 |

到港旅客转城轨
INTERCHANGE WITH INTERCITY RAILWAY FOR ARR PASSENGERS

| 1 | 迎客大厅 ARRIVAL HALL | 1F | ⇒ | 2 | 交通中心旅客大厅 GTC PASSENGER HALL | 1F | ⇒ | 3 | 扶梯下至城轨站厅 INTERCITY RAILWAY STATION | B1 |

到港旅客转大巴
INTERCHANGE WITH BUS FOR ARR PASSENGERS

| 1 | 迎客大厅 ARRIVAL HALL | 1F | ⇒ | 2 | 交通中心旅客大厅 GTC PASSENGER HALL | 1F | ⇒ | 3 | 交通中心巴士候车厅 GTC BUS WAITING HALL | 1F | ⇒ | 4 | 交通中心巴士站台 GTC BUS STATION | 1F |

到港旅客转小型车辆
INTERCHANGE WITH SMALL VEHICLE FOR ARR PASSENGERS

| 1 | 迎客大厅 ARRIVAL HALL | 1F | ⇒ | 2 | 交通中心旅客大厅 GTC PASSENGER HALL | 1F | ⇒ | 3 | 交通中心停车楼 GTC PARKING BUILDING | 1F |

到港旅客转出租车
INTERCHANGE WITH TAXI FOR ARR PASSENGERS

| 1 | 迎客大厅 ARRIVAL HALL | 1F | ⇒ | 2 | 迎客厅室外车道边 OUTDOOR CURBSIDE | 1F |

5.2.3 流程设计策略

1. 主流程旅客步行距离短。从车道边下车后，到最远登机口步行≤700m，步行时间＜9分钟。

2. 单一方向的旅客流程。因流线众多，为加强流线导向的清晰度，在流线设计上我们采取了单一方向、减少选择点的策略。旅客进入办票大厅后只有向前一个方向，不需做选择即到达安检厅，安检后开始选择登机口方向。到达旅客也是从各个指廊汇聚到主楼提取行李，国际和国内各有一个迎客口，统一的到达厅与交通中心连通，空间结构简单、方向引导非常清晰。

3. 国内流程采用混流设计。国内混流设计较出发、到达分流设计既减少了独立的到达楼层通道和服务机房面积，又共用了商业、卫生间、问询柜台等服务设施，还减少了登机口管理人员，减少了投资，提高了资源利用率。同时混流设计让国内转国内这一中转流程在平层中转，不需换层，提高了中转效率。

4. 国际流程仍采用分流设计。因海关和边防检查监管要求，采用了传统的出发、到达旅客分流的模式。

5. 国内/国际可切换的混合机位流程。混合机位对应候机指廊为国际出发层、国际到达层、国内混流层、设备机房与办公层等竖向四个功能楼层叠合设置，国际和国内旅客可同时分层候机，切换登机，提高了机位使用率。

6. 便捷的中转流程设计。中转流程是二号航站楼最重要的流程之一，其设计非常富有特点。国内转国内、国际转国内联程旅客到达、中转和候机全程在二层平层解决，国内转国际、国际转国际旅客流程经过主楼中转通道汇聚后，通过竖向交通引导至三层国际始发旅客联检口。中转流程检查共用始发、终到流程的检查场地资源，节约场地和人力资源，降低标识引导的复杂性，减少旅客选择点。

7. 国际国内并置的安检厅。在机场的检查流程上，二号航站楼最大的特点是将国际、国内安检并排设置在第一关，利用国际、国内高峰小时错峰的特点，对安检通道国际国内功能进行切换，解决特殊高峰时段（如春节）流量瓶颈问题。

8. 交通中心作为二号航站楼配套设施，其主要功能为供二号航站楼进出港的旅客与地面各种交通工具（地铁、城轨、大巴、出租车及私车）换乘的场所。

5.2.3 Flow Design Strategy

1. Short walking distance for major pedestrian flows.The walking distance from the curbside in the drop-off area to the farthest boarding gate is≤700m and requires a walking time＜9 minutes.

2. One-way passenger flow.The strategy of single direction with less options is adopted in flow design. Once entering the check-in hall, passengers have only one direction to proceed and can arrive at the security check hall with no other options. After the security check, they only need to select a boarding gate. The arrival passengers, likewise, converge from different piers into the main building to pick up baggage. One exit is provided respectively for domestic and international passengers, while the shared arrival hall is connected to GTC. The concise spatial layout contributes to extremely clear orientation.

3. Mixed domestic flows.Compared to separate design for departure and arrival flows, the mixed design for domestic flows not only reduces space for independent access and service room on the arrival floor, but also shares service facilities including retail, rest rooms, information desk etc., which in turn reduces the staff required at boarding gate, cut the investment and enhance the utilization efficiency of resources. The mixed-flow design also allows domestic-domestic transfer on the same floor to enhance transfer efficiency.

4. Separate international flows.Subject to customs and border inspection and relevant control requirements, the conventional design to separate departure flow from arrival one is adopted.

5. Swing stand flows. The swing stands serve waiting piers that are composed of four functional floors, i.e. international departure floor, international arrival floor, domestic mixed flow floor and MEP room and office floor. International and domestic passengers can wait on different floors at the same time thanks to the international-domestic convertible function of the swing stands that greatly improves stand efficiency.

6. Convenient transfer design.Transfer flow, as one of the most important flows in T2, has some very unique design. The arrival, transfer and waiting functions for domestic-domestic and international-domestic transfer passengers are all on F2. Domestic–international and international-international transfer passengers, after converging through the transfer access in the main building, are led by vertical transportation to the international departure passenger CIQ entrance on F3. The departure/arrival inspection site and resources are shared for the inspection of transfer passengers, which not only saves space and human resources but also reduce the complexity of signage system with less options available for passengers.

7. Parallel international and domestic security check halls.The most prominent feature of T2 in terms of airport inspection process is to place the international and domestic security check in parallel as the first pass. This takes advantage of the staggered peak hours of international and domestic security check to realize the conversion between international and domestic flights, hence a feasible solution to congestion of flows during special peak hours (such as the Spring Festival).

8. As a supporting facility of T2, the GTC mainly services as a place of interchange for passengers between T2 and various road transportation means (metro, intercity railway, bus, taxi and private cars).

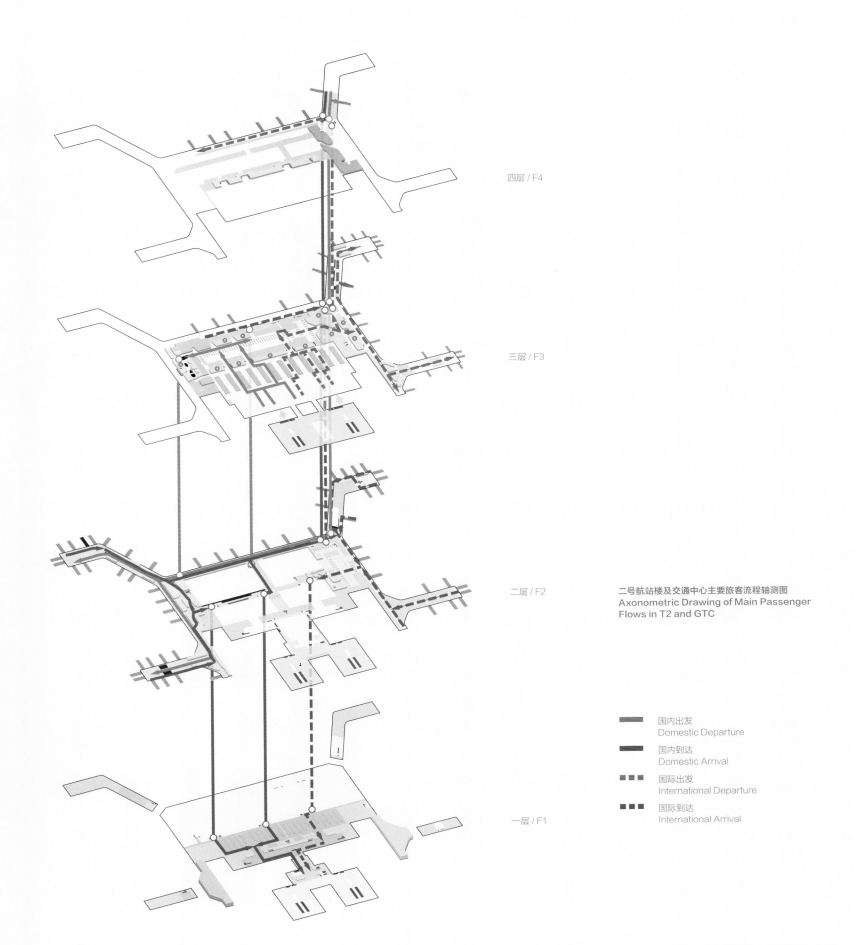

二号航站楼及交通中心主要旅客流程轴测图
Axonometric Drawing of Main Passenger Flows in T2 and GTC

5.3 功能配置 / 5.3 FUNCTIONAL CONFIGURATION

设施名称 Facility			设施数量(个) Quantity	备注 Remarks
值机柜台 Check-In Counter	国内 Domestic		142个人工柜台 142 manual counters	每个值机岛两侧均另设1个值班主任柜台、每个值机岛头设航空公司售票及服务柜台6个 Each check-in island has 1 chief on duty counter on each side of it and 6 ticket and service counters
			26个自助托运柜台 26 self-service baggage check-in counters	
			50台自助值机设备 50 self-service check-in devices	
			5个超大行李托运柜台 5 extra-large baggage check-in counters	
	国际 International		114个人工柜台 114 manual counters	每个值机岛两侧均另设1个值班主任柜台、每个值机岛头设航空公司售票及服务柜台6个 Each check-in island has 1 chief on duty counter on each side of it and 6 ticket and service counters
			26自助托运柜台 26 self-service baggage check-in counters	
			50台自助值机设备 50 self-service check-in devices	
			4个超大行李托运柜台 4 extra-large baggage check-in counters	
安检通道 Security Check Channel	国内 Domestic		48	含2条无障碍通道和1条员工通道 Including 2 accessible channels and 1 staff channel
	国际 International		22	含2条无障碍通道和1条员工通道 Including 2 accessible channels and 1 staff channel
检验检疫通道 Ciq Channel	出境 Departure		10条自助通道 10 self-service channels	
			6条人工查验通道 6 manual inspection channels	
			1个督导台 1 supervisory desk	
	入境 Arrival		8条自助通道 8 self-service channels	
			8条人工查验通道 8 manual inspection channels	
			1个督导台 1 supervisory desk	
海关通道 Customs Channel	出境 Departure		4条申报通道 4 claim channels	
			13条无申报通道数 13 no claim channels	
			1条员工通道 1 staff channel	
			1条外交礼遇通道 1 diplomatic courtesy channel	
	入境 Arrival		4条申报通道 4 claim channels	
			10条无申报通道数 10 no claim channels	
			1条员工通道 1 staff channel	
			1条外交礼遇通道 1 diplomatic courtesy channel	
边检通道 Frontier Inspection Channel	出境 Departure		30条人工通道 30 manual channels	
			20条自助通道 20 self-service channels	
			2条员工通道 2 staff channels	
	入境 Arrival		36条人工通道 36 manual channels	
			30条自助通道 30 self-service channels	
			2条员工通道 2 staff channels	
行李提取转盘 Baggage Claim Carousel	国内 Domestic		11	
	国际 International		10	

始发/目的地旅客功能配置
Functional Configuration for Departure/Arrival Passengers

名称 Description	细分类别 Detailed Category	流程及数量 Flows and quantity			
		国内转国内 Domestic-Domestic Transfer	国内转国际 Domestic-International Transfer	国际转国内 International-Domestic Transfer	国际转国际 International-International Transfer
中转柜台数量 Number Of Transfer Counters	无行李托运旅客 Passengers with no checked baggage	39个（国内指廊综合服务柜台通办D-D、D-I中转手续） 39 (comprehensive service counters for both D-D and D-I transfer procedures in domestic pier)	39个（国内指廊综合服务柜台通办D-D、D-I中转手续） 39 (comprehensive service counters for both D-D and D-I transfer procedures in domestic pier)	5个（联程旅客不提取行李，中转厅设置） 5 (no baggage claim for connecting flight passengers, provided in the transfer hall)	12个（直接过境旅客用） 12 (for direct transit passengers)
	有行李托运旅客 Passengers with checked baggage	6个（D-D与D-I共用） 6 (for both D-D and D-I transfers)	6个（D-D与D-I共用） 6 (for both D-D and D-I transfers)	13个（含一个超大托运） 13 (including one for extra-large baggage check-in)	
安检 Security Check	无行李托运旅客 Passengers with no checked baggage	联程旅客不需安检，混流厅内登机 No security check required and boarding in mixed-flow hall for connecting flight passengers	联程旅客不需安检，直接上至出境联检 No security check required and direct access upward to CIQ channel for connecting flight passengers	8条（联程旅客中转安检） 8 (security check for connecting flight passengers)	6条（直接过境，不提行李） 6 (direct transit without baggage claim)
	有行李托运旅客 Passengers with checked baggage	8条（D-D、D-I、I-D共用） 8 (for D-D, D-I, and I-D transfers)	8条（D-D、D-I、I-D共用） 8 (for D-D, D-I, and I-D transfers)	8条（D-D、D-I、I-D共用） 8 (for D-D, D-I, and I-D transfers)	
卫生检疫 Health Quarantine			走出境大流程查验场地 Use departure inspection area	人身检查走入境查验场地、随身行李在通关查验场地抽查 Physical security check at the arrival inspection area and random check of carry-on baggage at the customs clearance area	人身检查走入境查验场地、随身行李在通关查验场地抽查 Physical security check at the arrival inspection area and random check of carry-on baggage at the customs clearance area
海关 Customs	申报通道数 Number of claim channels		走出境大流程查验场地 Use departure inspection area	2个（联程旅客柜台） 2 (connecting flight passenger counters) 提取行李旅客走入境大流程查验场地 Passengers with baggage to claim use the arrival inspection channel	2
	无申报通道数 Number of no claim channels		走出境大流程查验场地 Use departure inspection area	5个（联程旅客柜台） 5 (connecting flight passenger counters) 提取行李旅客走入境大流程查验场地 Passengers with baggage to claim use the arrival inspection channel	2
	其他（员工、机组、礼遇等）通道数 Number of other (staff, crew, courtesy, etc.) channels		走出境大流程查验场地. Use departure inspection area		
边防 Frontier Defense	人工通道数 Number of manual channels		走出境大流程查验场地. Use departure inspection area	走入境大流程查验场地 Use arrival inspection area.	12个（直接过境，边防目前只监管，航空公司在使用柜台） 12 (direct transit, frontier defense only render supervision, airlines are using the counters) 6个（间接过境，过境免签旅客通道） 6 (indirect transit, TWOV channel)
	自助通道数 Number of self-service channels		走出境大流程查验场地 Use departure inspection area	走入境大流程查验场地 Use arrival inspection area.	
	其他（员工、机组、礼遇等）通道数 Number of other (staff, crew, courtesy, etc.) channels		走出境大流程查验场地 Use departure inspection area	走入境大流程查验场地 Use arrival inspection area.	

中转旅客功能配置
Functional Configuration for Transfer Passengers

候机区与登机口旅客功能配置
Functional Configuration for Waiting Areas & Boarding Gates

名称 Description	具体配置 Detailed Configuration
航延旅客服务设施 Flight-delayed Passenger Service Facilities	东五西五指廊设有航延旅客候机厅，分区设有一定数量航延服务柜台 Flight-delayed waiting halls in the East 5th Pier and West 5th Pier and some flight-delayed service counters by zone
航空公司柜台 Airline Counters	登机口柜台 Counters at boarding gates
有线电视 Cable Television	在旅客候机区结合座椅等其它设施布局有线电视 Cable televisions in the passenger waiting area in consideration of seats and other facilities

第6章

建筑设计
CHAPTER 6 ARCHITECTURE

6.1	设计构思	6.1	CONCEPTION
6.2	流动的造型	6.2	FLOWING FORM
6.3	外围护结构系统设计	6.3	BUILDING ENVELOP
6.4	主要空间衔接与室内设计	6.4	MAIN SPACE CONNECTIONS AND INTERIOR DESIGN
6.5	大空间照明设计	6.5	LARGE-SPACE LIGHTING
6.6	色彩体系设计	6.6	COLOR SYSTEM
6.7	标识系统设计	6.7	SIGNAGE SYSTEM
6.8	广告设计	6.8	ADVERTISEMENT
6.9	文化设计	6.9	CULTURAL FEATURES
6.10	主要旅客功能设施设计	6.10	FUNCTIONAL FACILITIES FOR PASSENGERS
6.11	景观设计	6.11	LANDSCAPE ARCHITECTURE
6.12	绿色建筑设计	6.12	GREEN BUILDING

6.1 设计构思

6.1 CONCEPTION

从一号航站楼开始,建筑师并没有刻意提出一个具体的符号或主题,只是强调一种现代、简洁、流畅、精致的设计风格。随着二号航站楼的创作,建筑师抽取白云元素,形成"云"概念。"云"概念构思包括多重内涵:

1. "云"概念,源自白云,与"白云国际机场"天然契合,也预示了旅客即将开始飞行于蓝天白云的梦幻旅程;
2. 在当今世界的语境中,"云"概念就是云端的连接,借此也表达了二号航站楼是旅客连接世界的地方;
3. "云"概念借鉴了当今最新科技云计算的理念,喻示着二号航站楼的设计与建设将集成当今时代的创新科技和理念,致力于为旅客带来良好的出行体验;
4. "云"概念与机场航站楼的功能特征非常吻合,飞行就是旅客群体的流动,交通建筑的效率与便捷非常重要。

The design of Terminal 1 (T1) emphasizes a modern, concise, smooth and exquisite design style instead of a specific symbol or theme. For Terminal 2 (T2), the concept of cloud is proposed based on the element of white cloud. The concept of cloud incorporates many meanings:

1. Inspired by white cloud, it naturally matches the name of Baiyun (i.e., white cloud) International Airport (Baiyun Airport), and signifies the fantastic journey above the white clouds in the blue sky;
2. In a context of the modern world, it means cloud connection, highlighting T2 as passengers' connection to the world;
3. It also means the modern technology of cloud computing, signifying the integration of modern innovative technologies and ideas in the design and development of T2 to bring the best travel experience for passengers;
4. It perfectly coincides with the functions of an airport terminal, as air flights represent the collective passenger flows, so the importance of efficiency and convenience of a transportation building should never be neglected.

一号航站楼正立面造型
Front Facade Form of T1

6.2 流动的造型

6.2 FLOWING FORM

从白云轻盈、漂浮、流动的特质出发，定义了二号航站楼设计的整体氛围和调性，塑造出流动的造型特色、轻盈的空间表现与柔和的空间格调，体现在以下几个方面：

1. 从"云"概念，提炼出连续云状拱形的造型母题；
2. "云"概念元素，成为二号航站楼建筑造型和空间的核心元素。连续多跨拱形的张拉膜雨蓬，与航站楼南立面檐口连续拱形曲面在造型上形成呼应，膜材漫反射的柔光映衬檐口的金属质感，形成层次丰富、光影变幻、连绵起伏的"云"意象；
3. 简洁且流畅的曲面形态整合了主楼与东西两翼的不同建筑体量，使之融合在一起，整体造型灵动、轻盈优雅、线条流畅、气势恢弘，为到访的旅客带来了个性化并具有亲和力的独特感受，简约并富有创意的手法与细节上的专注——这些都使得建筑给人的体验更加整体及流畅，同时也使航站楼更自然地与环境融合；
4. 保留了弧线形的主楼、Y形柱及张拉膜雨蓬这些白云机场特有的元素，与一号航站楼建筑造型和谐一致，共同建构了"双子星"航站楼的完整形象。

The light, floating and flowing features of clouds define the overall setting and tone of T2 design, which features flowing form, light spatial appearance and gentle spatial style as detailed below.

1. The main theme of the continuous cloud-shape arc is generated from the concept of cloud;
2. The concept of cloud serves as the core element in the architectural style of T2. The continuous multi-span arc tensile membrane canopy echoes with the continuous arc eave of south facade. The diffuse reflection of the membrane gently reflects the metal texture of the eave, creating a diversely-layered rolling cloud image with the change of light and shadows;
3. The concise and smooth curvy shape integrates the main building and its east and west wings that differ in volume. T2 boasts flexible form, light and elegant style, sweeping lines and magnificent appearance, impressing passengers with unique individuality and affinity. With concise and innovative approaches and great attention to details, the building not only provides a more holistic and smooth experience for passengers, but also realizes perfect harmony with the natural environment;
4. The unique elements of Baiyun Airport including curvy main building, Y-shape column and tensile membrane canopy are continued. By doing so, T2 maintains a harmonious and consistent architectural style with T1, presenting a holistic image of "Gemini" Terminal together with the latter.

二号航站楼正立面造型
Front Facade Form of T2

6.3 外围护结构系统设计

6.3.1 金属屋面系统

6.3.1.1 概述

二号航站楼金属屋面总面积约26万平方米。拱形非线性屋面造型连续流畅,主楼和连接楼相连一气呵成。全部采用外天沟以避免漏水。采用铝镁锰合金直立锁边屋面板系统,为加强金属屋面的抗风性能,采用三道措施进行加强。对应内部实用空间,在屋面布置采光天窗,主楼屋面布置一排整齐的条形天窗,指廊屋面的中间布置一条天窗。天窗造型简洁实用并兼顾屋面造型的美观,采用钢化夹胶中空玻璃(Low-E)。

6.3 BUILDING ENVELOP

6.3.1 Metal Roof System

6.3.1.1 System Description

T2's metal roof totals about 260,000m². The non-linear arc roof is continuous and smooth, with seamless transition between the main building and the connected buildings. The Project adopts outer gutter to avoid leakage, and Al-Mg-Mn alloy vertical edged roof board system with wind resistance reinforced through three measures. Skylights in the roof are provided corresponding to internal functional spaces, including a row of neat strip ones in the main building roof and a strip one in the middle of the pier roof. The concise and practical skylight design also contributes to the aesthetics of roof design. toughened laminated insulating glass (Low-E) is used for the skylights.

6.3.1.2 系统构成

1. 直立锁边金属屋面系统。铝镁锰合金屋面防水板、保温层、柔性防水层、纤维水泥板支撑层、岩棉支撑层、压型钢底板支撑层、主次檩条、主次檩托、屋面附加钢结构、屋面系统配件、屋面防跌落系统、屋面避雷系统等。

2. 屋面排水系统。屋面虹吸系统及重力排水系统,不锈钢天沟板、集水井、天沟龙骨、天沟附加钢结构。

3. 采光天窗系统。钢化夹胶中空玻璃(Low-E)、带导水槽铝合金副框、钢龙骨主受力构件、天窗避雷系统、气动开启天窗、室外铝单板收口及室内铝单板侧封板、铝合金收边配件等。

4. 屋面檐口系统。25mm厚蜂窝铝板(3mm厚铝单板);铝板铝合金副框、主次钢龙骨、檐口支撑钢结构构件、连接配件等。

6.3.1.3 支撑体系设计

二号航站楼平面外轮廓尺寸643m×295m,平面不规则,为避免过大的温度应力对结构的不利影响及满足抗震要求,对应钢结构屋架,通过设置温度缝(兼防震缝作用)将结构分割成数个较为规则的结构单元。支撑方式为在网架球节点上部设置屋面主檩,主檩条间距随网架网格确定,平面间距尺寸3m×3m;在主檩条上设置次檩条,次檩条间距不超过1.5m,局部加密为1m;次檩条上设置衬檩,衬檩上固定T码,T码用于直立锁边板的咬合固定。

6.3.1.4 构造防水

二号航站楼金属屋面设计执行《屋面工程技术规范》(GB50345-2012),主体屋面防水等级为1级。这就意味着金属屋面的构造设计无法完全按照一号航站楼模式,除了外层铝镁锰合金板防水层以外,内侧还须增加一道柔性防水卷材。

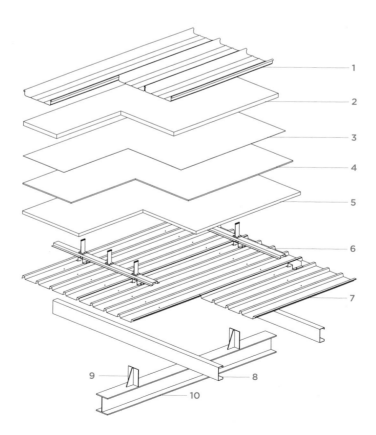

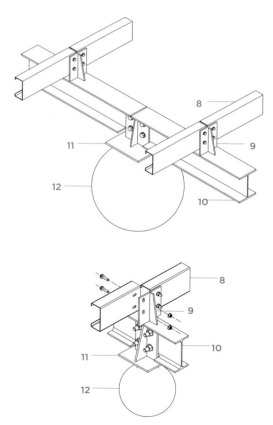

1. 屋面板(1mm厚氟碳预辊涂直立锁边铝镁锰合金板)
2. 保温层(两层50mm厚玻璃丝棉错缝铺设)
3. 防水层(1.2mm厚TPO防水卷材)
4. 支撑层(12mm厚纤维水泥板)
5. 支撑层(35mm厚岩棉层,下带加筋铝箔贴面)
6. 铝合金固定支座、几字形衬檩及衬檩支撑
7. 支撑层(0.6mm厚镀锌压型钢底板,肋高为35mm)
8. 次檩条
9. 主檩及次檩托板
10. 主檩条
11. 网架球支托
12. 网架球

1. Roof Board (vertical edged al-mg-mn alloy board with 1mm fluorocarbon pre-roll coating)
2. Insulation Layer (50 mm two-layer glass wool in staggered-joint paving)
3. Waterproof Layer (1.2 mm tpo waterproof roll)
4. Supporting Layer (12 mm fiber cement board)
5. Supporting Layer (35 mm rock wool with reinforced aluminum foil veneer underneath)
6. Aluminum Alloy Fixed Mount, Ω-Shaped Lining Purlin and Support
7. Supporting Layer (0.6 mm galvanized profiled steel base with a rib of 35 mm high)
8. Secondary Purlin
9. Primary and Secondary Purlin Supporting Board
10. Primary Purlin
11. Grid Spherical Support
12. Grid Sphere

屋面构造图
Roof Construction Plan

屋面造型1
Roof Shape 1

屋面造型2
Roof Shape 2

屋面檐口及天窗系统
Roof Cornice and Skylight System

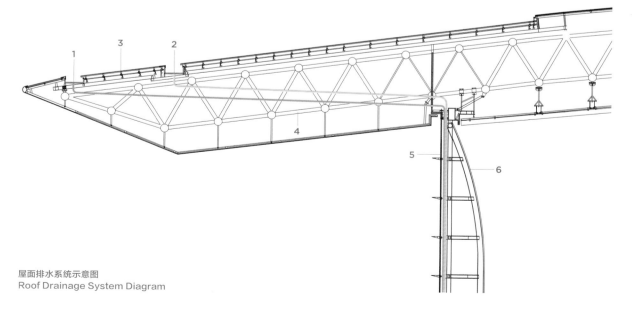

1. 重力雨水斗　　1. Gravity Roof Outlet
2. 虹吸雨水斗　　2. Siphonic Roof Outlet
3. 直立锁边金属屋面　3. Vertical Edged Metal Roof
4. 带肋钢网架　　4. Ribbed Steel Grid
5. 玻璃幕墙　　　5. Glass Curtain Wall
6. 钢桁架抗风柱　6. Streel Truss Wind Resistant Column
7. 不锈钢板天沟　7. Stainless Steel Gutter
8. 气动排烟天窗　8. Pneumatic smoke exhaust skylight
9. 采光天窗　　　9. Skylight
10. 网架球　　　10. Grid Sphere
11. 网架球支托　11. Grid Spherical Support
12. 连接板　　　12. Connecting Board
13. 连接立柱　　13. Connecting Column
14. 主檩条　　　14. Primary Purlin

屋面排水系统示意图
Roof Drainage System Diagram

6.3.1.2 System Composition
1. Vertical edged metal roof system. Al-Mg-Mn alloy roof waterproof boards, insulation layer, flexible waterproofing layer, fiber cement board supporting layer, rock wool supporting layer, profiled steel base plate supporting layer, primary and secondary purlins, primary and secondary purlin hangers, additional roof steel structure, roof system fittings, roof fall protection system, roof lightning protection system etc..
2. Roof drainage system. Roof siphon system and gravity drainage system, stainless steel gutter plate, sump pit, gutter joist, additional steel structure for gutter.
3. Skylight system. Tempered laminated insulating glass (Low-E), secondary aluminum alloy frame with water chute, main stress member of steel joist, skylight lightning protection system, pneumatically operable skylight, exterior aluminum panel edging and interior aluminum panel side seal, aluminum alloy close-up fittings etc.
4. Roof eave system. 25mm thick honeycomb aluminum plates (3mm thick aluminum panels); secondary aluminum alloy frame for aluminum plates, primary and secondary steel joists, eave supporting steel structure members, connecting pieces etc..

6.3.1.3 Supporting System
T2 has an irregular planar shape with a contour of 643 m×295m. To avoid negative impact of high temperature stress on structure and comply with seismic requirements, the steel roof truss is divided into relatively regular structural units by temperature joint (also seismic joint). The supporting measures include providing primary roof purlins on top of the grid spherical nodes, at an interval subject to gridding and a planar spacing of 3 m×3 m; providing secondary purlins on primary purlins at an interval of no more than 1.5 m (1m at some positions); providing linear purlins on secondary purlins, and fixing T-shape supports on linear purlins to secure vertical edging plates.

6.3.1.4 Constructional Waterproofing
Metal roof of T2 follows the *Technical Code for Roof Engineering* (GB 50345-2012), with Grade 1 waterproofing for the main roof. This means that T2's metal roof constructional design cannot completely follow the approach of T1. In addition to exterior layer of Al-Mg-Mn alloy waterproof board, an interior layer of flexible waterproof roll must be added.

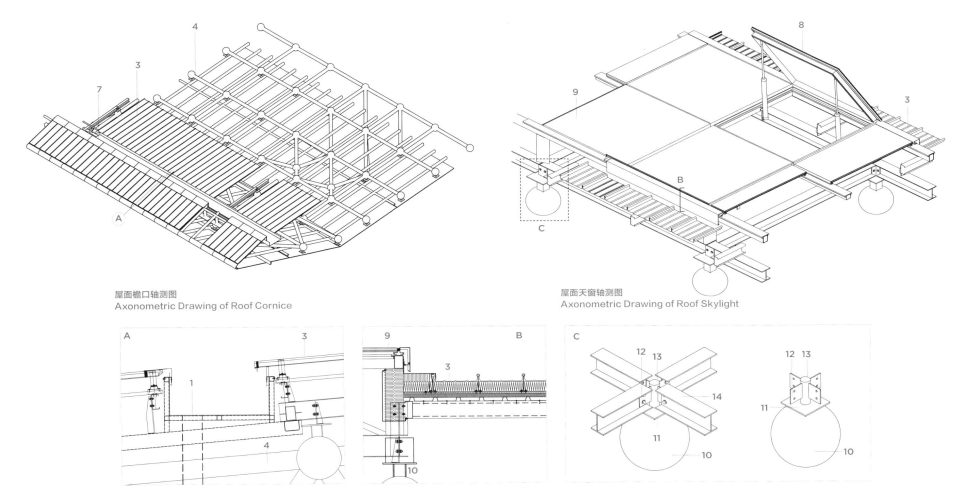

屋面檐口轴测图
Axonometric Drawing of Roof Cornice

屋面天窗轴测图
Axonometric Drawing of Roof Skylight

6.3.1.5 抗风设计

1. 风荷载取值

（1）结构设计使用年限
建筑的设计基准期为50年；
航站楼的设计使用年限在承载力及正常使用情况下为50年；
耐久性下重要构件为100年，次要构件为50年。

（2）建筑安全等级
建筑物安全等级为一级，重要性系数$\gamma_0=1.1$。

2. 檩条排布。 根据风洞试验报告，取各区域边缘处最大风压值进行计算，再将计算结果放大2.0倍，用于屋面系统的抗风取值，得出安全稳妥的檩条布置方式。

3. 抗风系统。 为了保证金属屋面系统的抗风安全性，在金属屋面设计完成后，又为它制定了一套完整的抗风系统。整套系统采用三道措施加强屋面抗风性。

（1）屋面板板尾是最危险的位置，没有封口铝板保护。板尾增加两排抗风夹，抗风夹固定于屋面板最边缘两道T码上，抗风夹用10mm不锈钢棒连接成整体。

（2）屋面板板头有屋脊或是天窗收口保护。板头也增加两排抗风夹，抗风夹固定于离天窗最近的两道T码上，抗风夹用10mm不锈钢棒连接成整体，作为第二道保险。

（3）顺板方向，增加一条10mm不锈钢索，每36m一根，两端与檐口龙骨固接（钢索不需张拉受力）。与不锈钢棒形成纵横交错。万一整块板掀起，将被索拴在屋面上，作为第三道保险。

6.3.1.5 Wind Resistance

1. Wind Load Values

(1) Designed Structural Service Life
Design reference period of the building is 50 years;
The designed service life of the terminal is 50 years within designated bearing capacity under normal use;
The durability of key elements is 100 years, and secondary elements is 50 years.

(2) Building Safety Grade
Building safety is Grade I, and importance factor $\gamma_0=1.1$.

2. Purlin Layout. Based on wind tunnel report, maximum wind pressure values at the edges of all areas are adopted for calculation, and the calculated results are doubled to get the wind resistance value for roof system and eventually the safest and most secure purlin arrangement.

3. Wind Resistance System. To ensure safe wind resistance for the metal roof system, after metal roof design is completed, a complete set of wind resistance system is worked out with three measures taken to strengthen wind resistance for the roof.

(1) Roof plate tail, without the protection of sealing aluminum plate, is the most dangerous position. Two rows of wind resistance clips are added to the plate tail and secured to the two rows of T-shape supports along the outer edge of the roof plate. Wind resistance clips are connected by 10 mm stainless steel rods into a whole.

(2) Roof plate nose is protected by roof ridge or skylight close-up. The nose also has two rows of wind resistance clips secured to the two rows of T-shape supports closest to skylight. Wind resistance clips are connected by 10 mm stainless steel rods into a whole as the second protective measure.

(3) Along plate direction, a 10mm stainless steel rope is installed every 36m, with both ends secured to eave joists (no tensile stress on the steel rope). With steel ropes crisscrossing steel rods, even if the entire plate is lifted up, it would still be secured to roof by the steel rope. This serves as the third protective measure.

屋面天窗及抗风系统
Skylight System and Roof Wind-resistant System

檩条布置方式
Purlin Layout

序号 Number	措施项目 Measures		理论设置 Theoretical Setting	实际设置 Actual Setting
1	基本风压值 Basic Wind pressure Value		0.60 kn/m²	0.69 kn/m²
2	檩条间距 Purlin Interval	中间区 Intermediate Zone		1.5m
3		边缘区 Marginal Zone		1.0m
4	边缘区宽度 Width of Marginal Zone	主楼 Main Building		13.5m
5		指廊 Pier		7.5m

6.3.2 幕墙系统

6.3.2.1 概述

二号航站楼玻璃幕墙总面积约10万平方米，其中最大尺度幕墙为主楼正立面，总面宽约430m，最大高度36m。与点支式玻璃幕墙的一号航站楼不同，二号航站楼采用横明竖隐的玻璃幕墙系统，玻璃板块尺寸为3m×2.25m，高度分格与建筑层高模数2.25m相同，宽度分格与建筑平面模数相同。二次结构采用立体钢桁架为主受力构件，横向铝合金横梁为抗风构件，竖向吊杆为竖向承重构件。铝合金横梁也起到水平遮阳作用，采用挤压铝型材。竖向吊杆藏于玻璃接缝中，整个幕墙单元具有通透感。立体钢桁架结构无需水平侧向稳定杆，幕墙整体感强、通透美观。设备管线与幕墙二次结构相结合，将管线埋藏于U型结构内，侧面再用铝扣板遮挡。为配合消防排烟开启扇角度达到60°的需要，采用可以大角度开启的气动排烟窗。考虑夏热冬暖地区的节能需要，西侧幕墙外侧设置机翼型电动可调遮阳百叶。有0°、30°、60°和90°四个角度。百叶选用铝合金微孔板，可以达到遮阳且透影的效果。

6.3.2.2 支撑体系设计

建筑结构安全等级一级，设计使用年限50年。按《建筑抗震设计规范》（GB50011-2010），二号航站楼抗震设防烈度为6度，抗震构造设防按7度。场地类别为II类场地。

主楼、各指廊及相关连廊的幕墙工程主要采用横明竖隐玻璃幕墙系统和铝板幕墙形式。主楼玻璃幕墙采用竖向三角钢桁架，每12m一榀；连廊与指廊采每9m一榀。竖向钢桁架和钢桁架之间采用横向铝横梁共同承受风荷载，竖向吊索仅承受幕墙自重的结构形式，通过横向铝横梁来实现横明竖隐的立面效果。

6.3.2.3 功能设计

1. 主要性能指标。按照《建筑幕墙》（GB21086-2007）等相关规范，结合材料与分隔，通过计算，确定本工程幕墙相关性能标准：风压变形性能为2级，水密性应达到2级，气密性能分级为3级，平面变形性能为5级，传热系数达到6级，遮阳系数达到6级，空气声隔声性能等级为3级，耐撞击性能为2级。

2. 防火设计。幕墙的防火等级为一级，1小时防火时限。对于跨层的幕墙，在层间梁处设置难燃防火层，并避免同一玻璃板块跨越两个防火分区，具体做法为1.5mm镀锌钢板槽内填100mm厚防火岩棉封闭封堵。无窗间墙的幕墙，在每层楼板外沿设置耐火极限不低于1h、高度不低于0.8m的不燃实体裙墙。幕墙与每层楼板、隔墙处的缝隙采用防火材料封堵，防火岩棉的密度不小于110kg/m³，厚度不小于100mm。承托板与主体结构、幕墙结构及承托板之间的缝隙填充防火密封胶。

3. 防雷设计。按照《建筑物防雷设计规范》（GB50057-94）中防雷分类等级的第二类防雷标准进行防雷设计。将幕墙金属部件和龙骨连通，同时与主体防雷网进行连通。

4. 防腐蚀设计。在两种不同金属材料接触的部位设置防腐蚀橡胶垫，防止电化学腐蚀。铝型材表面做氧化或氟碳喷涂处理，钢件都进行镀锌、氟碳喷涂或刷防腐漆处理。最大限度地采用螺栓连接，防止大面积现场烧焊对防腐膜层的破坏。非外露钢材的表面热浸锌处理，钢材厚度小于5mm的镀锌层厚度≥65μm，钢材厚度大于5mm的镀锌层厚度≥85μm。外露部分钢材表面处理为表面氟碳喷涂。

5. 防噪声设计。为消除铝横梁由于温度变化产生的噪声，在立柱与横梁连接处加橡胶垫片。金属与金属间可能存在相对移动的地方增加防噪声胶条。

6. 防结露设计。本项目幕墙的玻璃为中空LOW-E玻璃，金属板为蜂窝铝板，经过计算，采用以上措施可以使维护结构的总热阻大于产生空气露点温度的热阻，避免了结露的产生。

7. 采光、防止光污染设计。玻璃幕墙的可见光透射比不低于0.4，玻璃幕墙均采用反射比不大于0.3的玻璃。

8. 清洁维护设计。在屋面檐口下沿预留幕墙清洗吊挂孔，方便幕墙清洁。

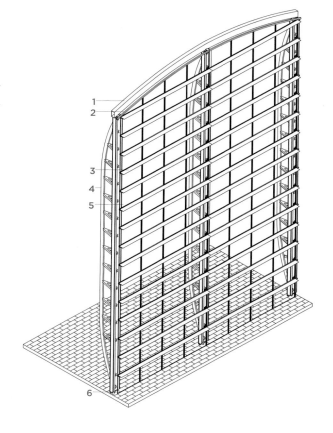

主楼标准玻璃幕墙段
Typical Glass Curtain Wall of Main Building

1. 幕墙顶部弧形箱型梁
2. 弧形铝合金横梁
3. 铝合金横梁
4. 钢桁架抗风柱
5. 中空夹胶钢化彩釉玻璃
6. U型钢收边底槽

1. Arc Box-Shaped Beam on Top of Curtain Wall
2. Arc Aluminum Alloy Beam
3. Aluminum Alloy Beam
4. Steel Truss Wind-Resistant Column
5. Insulated Laminated and Tempered Fritted Glass
6. U-steel Bottom Cut

迎客厅玻璃幕墙内景
Internal View of Glass Curtain Wall of Arrival Hall

6.3.2 Curtain Wall System

6.3.2.1 Description

The total area of glass curtain wall of T2 is about 100,000m2, with the largest scale on the front facade of the main building, measuring about 430m wide and 36m high at the highest point. Different from T1's point-supported glass curtain wall, T2 adopts a glass curtain wall system with horizontally exposed and vertically hidden frames. The glass panel is sized 3 m X2.25 m, with height gridding the same as the floor height module of 2.25m, and width gridding the same as that of planar module. The secondary structure adopts multi-level steel truss as the main load-bearing member, transverse aluminum alloy beam as wind resistant member and vertical hanger as vertical load-bearing member. Extruded aluminum alloy beams are adopted for horizontal sunshading. Vertical hangers are hidden in glass joints, making the entire curtain wall a transparent unit. The multi-level steel truss structure requires no horizontal lateral bars as stabilizer, making the curtain wall highly integrated, transparent and aesthetic. MEP pipelines are integrated with the secondary structure of curtain wall, with pipelines buried in U-shape structure, and the flank covered by aluminum panels. As fire smoke exhaust requires the operable sash to be open at 60 degrees, pneumatic smoke exhaust windows with big operable angle are used. As Guangzhou is hot in summer and warm in winter, wing type adjustable electric louvers that support operable angles of 0°, 30°, 60° and 90°are installed on the western curtain wall for energy efficiency. Micro-perforated aluminum louvers are adopted for both sunshade and visibility.

6.3.2.2 Supporting System

Building structural safety is Grade I and the designed service life is 50 years. In accordance with the *Code for Seismic Design of Buildings (GB 50011-2010)*, seismic fortification intensity of T2 is 6, and seismic structure fortification intensity of T2 is 7, with the classification of site falling in the scope of Class II.

The main building, piers and corridors are mainly designed with horizontally exposed and vertically hidden curtain wall system and aluminum plate curtain wall. Vertical triangular steel truss is used for glass curtain wall of the main building, each 12m, and for corridor and pier, each 9m. Horizontal aluminum beams are designed between vertical steel trusses to jointly bear wind load, while vertical slings only bear the dead weight of the curtain wall, with the horizontally exposed and vertically hidden facade effect realized by horizontal aluminum beams.

6.3.2.3 Functional Design

1. Major performance indicators. In accordance with the *Curtain Wall for Building (GB21086-2007)* etc., and through calculations in consideration of material and partition, relevant performance indicators for curtain wall of this Project is defined: deformation under wind pressure is Level 2,water tightness Level 2,air tightness Level 3,planar deformation Level 5,heat transfer coefficient Level 6,sun-shading co-efficient Level 6; air sound insulation Level 3,anti-collision Level 2.

2. Fire design. Curtain wall fire rating is Level I, and fire resistance rating is 1 hour. For curtain wall crossing floors, fire resistance layer is designed at beams between floors, and attention is paid to avoid one same glass panel crossing two different fire zones. Specifically, the design adopts 1.5 mm galvanized steel slot filled with 100 mm thick fireproof rock wool. Curtain wall without pier between two windows is provided with non-combustible solid wainscot of no less than 0.8 m high with fire resistance of no less than 1h along the outer slab edge of each floor. The gaps between curtain wall and the slab and partition wall of each floor are filled with fireproof rock wool of no less than 110 kg/m³ in density and 100 mm in thickness. The gaps between supporting plate and the main structure, and between curtain wall structure and supporting plate are filled with fireproof sealant.

3. Lightning protection. Lightening protection design follows Class II lightning protection standards in the *Code for Lightning Protection Design of Buildings (GB50057-94)*. Curtain wall metal parts are connected with joists and the

玻璃幕墙外立面
Glass Curtain Wall Facade

lightning protection network of the main building.

4. Anti-corrosion design. Anti-corrosion rubber pads are installed at the contacts between different metals to avoid electrochemical corrosion. Aluminum surfaces are subject to oxidation or PVDF coating, while steel pieces are subject to galvanization, PVDF or anticorrosion coating. Bolt connections are implemented to the greatest extent to avoid damage to anticorrosion layer caused by large areas of on-site welding. Unexposed steel surfaces are subject to hot-dipped galvanization, with zinc coating ≥ 65Qm thick for steel materials less than 5 mm thick, and zinc coating ≥ 85Qm thick for steel materials more than 5 mm thick. Exposed steel surfaces are subject to PVDF coating.

5. Noise control. To eliminate noise of aluminum beam caused by temperature change, rubber pads are installed between column and beam, and soundproof strips are installed at metal connections where relative movement is likely to occur.

6. Anti-condensation. Curtain wall of the Project adopts LOW-E insulating glass and honeycombed aluminum plates. Calculation shows that the above measures can make the total thermal resistance of envelop structure greater than that of air dew-point temperature, thus avoiding condensation.

7. Daylighting and light pollution control. Visible light transmission of the glass curtain wall is no less than 0.4, and reflectance ratio of the glass for curtain wall is no more than 0.3.

8. Maintenance. Hoist holes are reserved below roof eave for convenient cleaning of the curtain wall.

6.3.2.4 构造选型

二号航站楼采用了框式幕墙中的横明竖隐的结构体系。玻璃幕墙以大分格、大跨度玻璃为主，玻璃板块尺寸3m×2.25m。二次结构采用立体钢桁架为主受力构件，横向铝合金横梁为抗风构件，竖向吊杆为竖向承重构件。

6.3.2.5 节点设计

二号航站楼采用竖向钢桁架作为立挺，支撑顶部钢结构大横梁，玻璃缝隙之间隐藏竖向不锈钢拉杆外加铝盖板，拉杆上部吊挂于顶部大横梁上，拉杆负责承托幕墙横梁，横梁既是建筑水平装饰构件，又作为幕墙结构体系中水平抗风杆件。横梁采用挤压铝型材，外观平整，截面细小，截面高度小，12m为整根型材。

6.3.2.6 设计创新

二号航站楼的幕墙设计，采用了成熟的横明竖隐的框式幕墙系统，具有多项技术创新：

1. 采用横明竖隐的玻璃幕墙。玻璃板块尺寸为3m×2.25m，高度分格与层高模数2.25m相同，主要层高包括4.5m、9m、11.25m，宽度分格与平面模数相同，主要柱网为9m、18m、36m。
2. 设备管线与幕墙二次结构相结合。将管线埋藏于U型结构内，侧面再用铝扣板遮挡。为了隐藏虹吸雨水管及强弱电管线，立挺设计之初就在钢结构与幕墙之间预留凹槽，并在混凝土结构板对应位置预留虹吸雨水套管，并结合特殊的柱脚设计，使所有管线隐藏于幕墙中。

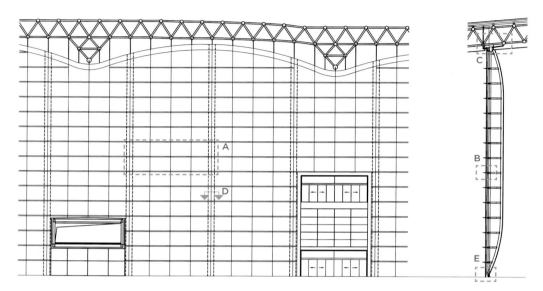

主楼标准玻璃幕墙段立剖面
Elevation and Section of Typical Glass Curtain Wall of Main Building

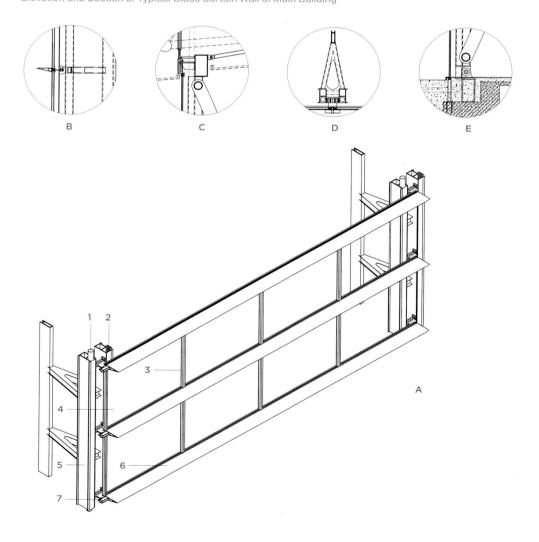

玻璃幕墙标准单元
Typical Unit of Glass Curtain Wall

1. 落水管
2. 电气管线
3. 竖向不锈钢拉杆外加铝盖板
4. 中空夹胶钢化彩釉玻璃
5. 钢桁架抗风柱
6. 铝合金横梁
7. 铝型材套管

1. Downpipe
2. Electrical Pipeline
3. Vertical SS Rod with Aluminum Capping
4. Insulated, Laminated and Tempered Gritted Glass
5. Steel Truss Wind-Resistant Column
6. Aluminum Alloy Beam
7. Aluminum Profile Sleeve

6.3.2.4 Structural System

T2 adopts horizontally exposed and vertically hidden frame curtain wall. The glass curtain wall mainly consists of glass panels of large grid and long span in the size of 3 mX2.25 m. For the secondary structure, multi-level steel truss is adopted as main load bearing member, horizontal aluminum alloy beam as wind resistance member and vertical hanger as vertical load bearing member.

6.3.2.5 Node Design

T2 adopts vertical steel trusses as mullions to support the large steel beams on top. Hidden in glass joints are vertical stainless steel rods with aluminum capping, the upper part of which is attached to the large beams at the top. The rods are responsible for supporting curtain wall beams, which serve as both horizontal decoration members for the building and horizontal wind resistant pieces in curtain wall structural system. The beams are made of extruded aluminum profile, with flat surface and small section height. Each beam is 12m long.

6.3.2.6 Design Innovations

T2 adopts the mature system of horizontally exposed and vertically hidden frame curtain wall that incorporates multiple innovative technologies:

1. Glass curtain wall with horizontally exposed and vertically hidden frames. Glass panels are 3mX2.25m in size, with height gridding the same as floor height modulus of 2.25m (major floor heights include 4.5m, 9m and 11.25m) and width gridding the same as planar modulus (main column grids includes 9m, 18m and 36m).
2. MEP pipelines integrated with the secondary structure of curtain wall. The pipelines are buried in U-shape structure, with the flank covered by aluminum panels. To conceal siphon rainwater pipelines, HV and ELV pipelines, slots between steel structure and curtain wall had been decided early at the beginning of design, and siphon rainwater casings had been reserved at positions corresponding to concrete structural slabs with special column footing design to hide all the pipelines in the curtain wall.

玻璃幕墙节点
Glass Curtain Wall Nodes

1. 落水管
2. 电气管线
3. 钢桁架抗风柱
4. 铝型材转接件
5. U型钢转接件
6. 铝型材套管
7. 中空夹胶钢化彩釉玻璃
8. 竖向不锈钢拉杆外加铝盖板
9. 铝合金横梁

1. Downpipe
2. Electrical Pipeline
3. Steel Truss Wind-Resistant Column
4. Aluminum Profile U-steel
5. U-steel Adaptor
6. Aluminum Profile Sleeve
7. Insulated, Laminated and Tempered Gritted Glass
8. Vertical SS Rod with Aluminum Capping
9. Aluminum Alloy Beam

3. 幕墙的横梁设计。国内首创采用特殊截面横梁设计+12m通长大横梁+9m通长曲线横梁（500mm截面）的铝型材造型横梁作为结构受力构件。

4. 结合幕墙设置机翼型可调节电动遮阳百叶。广州地区夏热冬暖，西侧的遮阳措施非常重要，本工程在幕墙外侧布置了机翼型可调节电动遮阳百叶，有0°、30°、60°和90°四个角度。百叶选用铝合金微孔板，可以达到遮阳且透影的效果。

5. 气动排烟窗的设计。按照消防排烟要求，排烟窗开启扇的角度需达到60°，因此采用可以大角度开启的气动排烟窗，运用了特殊的防失效设计，保证排烟窗在断电、断气、断消防信号或者三者全断情况下，依然可以开启排烟。

3. Curtain wall beam design. The design adopts special section beam design + 12m full-length big beam + 9m full-length curvy beam (500mm section), first of its kind in China, as structural load-bearing member.

4. Wing type adjustable electric louvers on curtain wall. Guangzhou is hot in summer and warm in winter, so shading measures on the west facade are very important. T2 is equipped with wing type adjustable electric louvers outside the curtain wall with operable angles of 0°, 30°, 60° and 90°. Micro-perforated aluminum louvers are adopted for both sunshade and visibility.

5. Pneumatic smoke exhaust windows. As fire smoke exhaust requires the operable angle of smoke exhaust operable sashes to reach 60°degrees, pneumatic smoke exhaust windows with large operable angle are adopted. Special anti-failure design is implemented to ensure normal smoke exhaust in case of failure of power, gas or fire signal or all of them.

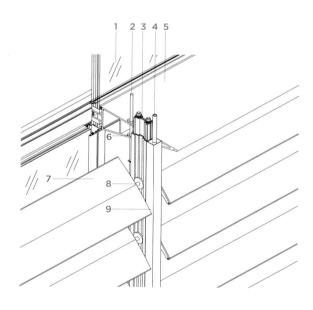

1. 中空夹胶钢化彩釉玻璃
2. 电动遮阳叶片驱动连杆
3. 铝合金套芯
4. 吊杆
5. 铝合金横梁
6. 横梁长圆孔
7. 耐色光电动遮阳百叶
8. 叶片转动轴
9. 遮阳铝通立柱

1. Insulated, Laminated and Tempered Gritted Glass
2. Connecting Rod Driver of Electric Louvre Blade
3. Aluminum Alloy Core
4. Boom
5. Aluminum Alloy Beam
6. Slotted Hole of Beam
7. Light-Resistant Electric Louvers
8. Blade Rotary Shaft
9. Sun-Shading Aluminum Column

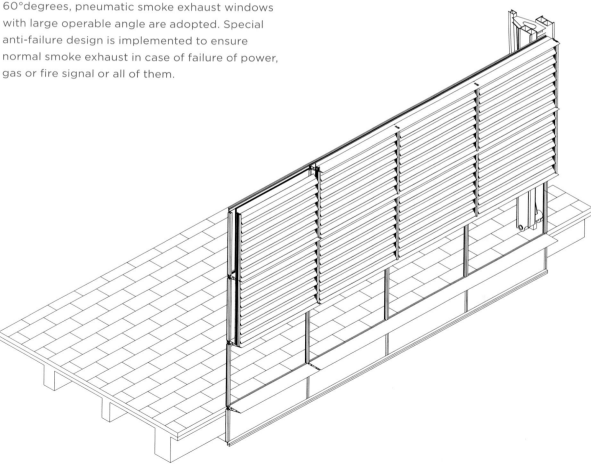

可调节电动遮阳百叶与幕墙结合标准单元
Typical Integration Unit of Adjustable Electric Louvers and Curtain Wall

幕墙结合遮阳设计
Curtain Wall combined with Sunshading

6.4 主要空间衔接与室内设计
6.4 MAIN SPACE CONNECTIONS AND INTERIOR DESIGN

6.4.1 主要空间衔接
6.4.1 Main Space Connections

1. 交通中心停车楼	12. 行李提取厅	
2. 屋顶绿化停车场	13. 国内到达走廊	
3. 交通中心旅客大厅	14. 主楼四层餐饮平台	
4. 交通中心过厅	15. 安检大厅	
5. 大巴隧道	16. 行李分拣机房	
6. 的士隧道	17. 岭南花园	
7. 出发层高架桥	18. 国内混流候机厅	
8. 迎客大厅	19. 国际到达走廊	
9. 北进场隧道	20. 国际出发候机厅	
10. 地下管廊	21. 登机桥	
11. 办票大厅		

1. GTC Parking Building
2. Roof Greening Parking Lot
3. GTC Passenger Hall
4. International Baggage Sorting Machine Room
5. Bus Tunnel
6. Taxi Tunnel
7. Departure Floor Viaduct
8. Arrival Hall
9. North Approach Tunnel
10. Underground Utility Tunnel
11. Check-in Hall
12. Baggage Claim Hall
13. Domestic Arrival Channel
14. F4 F&B Platform of Main Building
15. Security Check Hall
16. Baggage Sorting Machine Room
17. Lingnan Gardens
18. Domestic Mixed-flow Waiting Hall
19. International Arrival Channel
20. International Departure Waiting Hall
21. Boarding bridge

6.4.2 室内设计

二号航站楼室内空间设计延续"云"概念，由建筑外观造型到室内空间，由外而内，犹如行云流水，层层递进，一气呵成。从出发车道边到办票大厅，由三条通廊一直贯穿到安检大厅，最终延伸到候机指廊，从室外城市空间到室内公共空间，"云"概念元素在建筑造型与空间中流动转换，节奏灵动流畅，塑造出具有原创性的二号航站楼空间形象；流动、轻盈的空间表现、黑白灰柔和的空间调子，体现在以下几个方面：

1、具有强烈动感的波浪形旋转渐变吊顶，体现空间的流动性，并有良好的导向性。
2、楔形线条及体量对流动性的表达，设计上引入交汇的楔形线及体量，有别于正交形体，由于不平行的楔形线在透视上的错觉，在空间上具有不稳定感，从而带来很好的流动感。
3、引入自然光线，光影赋予空间流动性，同时自然地引领旅客，让旅客身心舒畅。
4、以共享空间的手法营造重要空间节点，衔接交通空间与商业空间，也充满流动性。

6.4.2 Interior Design

The interior space design of T2 continues the "cloud" concept. The interior spaces, as extension of exterior building forms, progress from the outer to the inner parts freely like flowing water. From departure curbside to check-in hall, three corridors run through the security check hall until the waiting pier. From urban space outside to public space inside, the concept of cloud flows and changes throughout building forms and spaces in a rhythmic and fluent manner, shaping up the original spatial image of T2. The lightsome flowing spaces and gentle spatial tune featuring a mix of black, white and grey are reflected in the following aspects:

1. Highly dynamic wavy suspended ceiling in gradual rotation reflects the flow the space and presents a good sense of orientation.
2. Wedge lines and shapes for the expression of fluidity. Intersecting wedge lines and shapes are introduced in design. Different from orthogonal shapes, the illusion of unparalleled wedge lines in perspective give a sense of instability in space and therefore good fluidity.
3. Introduction of natural light. Natural light and shadows create fluidity in spaces, and at the same time, provide natural guidance and comfortable experience for passengers.
4. Major spatial nodes. Major spatial nodes are created through the approach of shared spaces to connect and add fluidity to transportation and commercial spaces.

办票大厅
Check-in Hall

办票大厅
Check-in Hall

采光天窗及旋转天花叶片剖面
Section of Skylight and Rotating Ceiling Blade

1. 办票大厅
2. 旋转天花叶片
3. 采光天窗
4. 直立锁边金属屋面
5. 带肋钢网架
6. 办票岛
7. 钢管混凝土柱

1. Check-in Hall
2. Rotating Ceiling Blade
3. Skylight
4. Vertical Edged Metal Roof
5. Ribbed Steel Grid
6. Check-in Island
7. Concrete Filled Steel Tubular Column

办票大厅
Check-in Hall

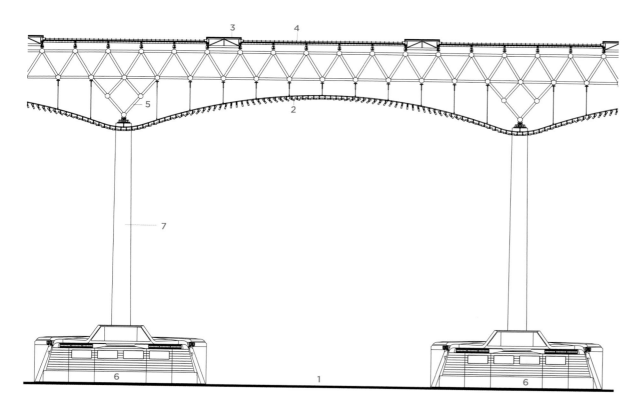

办票大厅商业及四层餐饮平台
Commercial and F4 F&B Platform of Check-in Hall

办票大厅往安检大厅通道
Passage between Check-in Hall and Security Check Hall

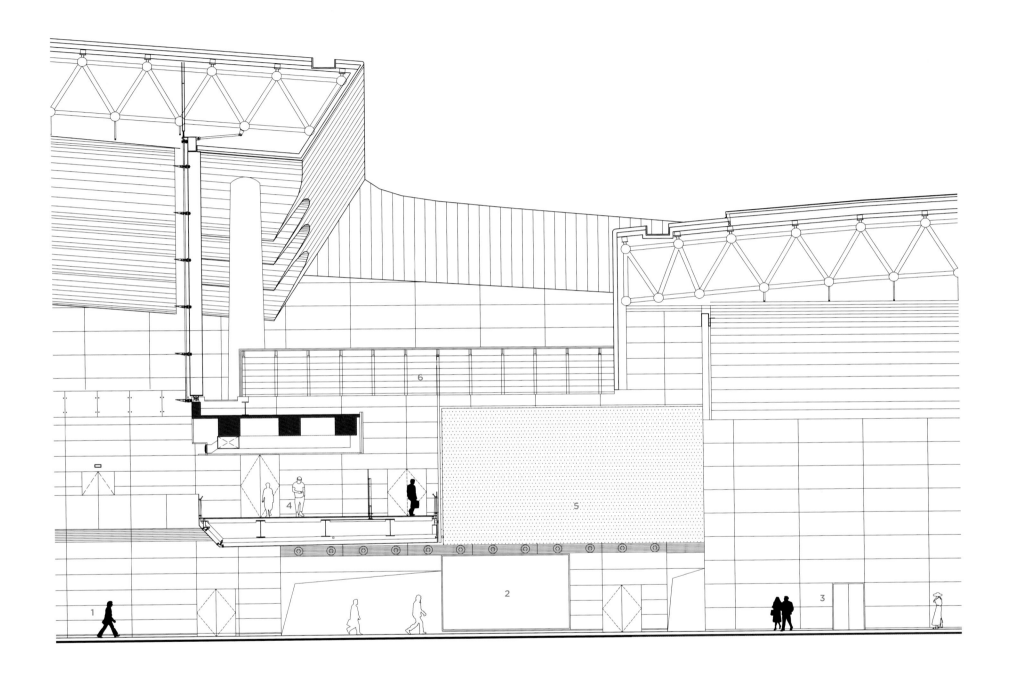

办票大厅往安检大厅通道剖面
Section of Passage between Check-in Hall and Security Check Hall

1. 办票大厅
2. 办票大厅往安检大厅通道
3. 安检大厅
4. 主楼四层餐饮平台
5. 绿化墙面
6. 采光天窗

1. Check-in Hall
2. Passage between Check-in Hall and Security Check Hall
3. Security Check Hall
4. F4 F&B Platform of Main Building
5. Greening Wall
6. Skylight

出境边检大厅
Departure Frontier Inspection Hall

安检大厅
Security Check Hall

国内商业区
Domestic Commercial Area

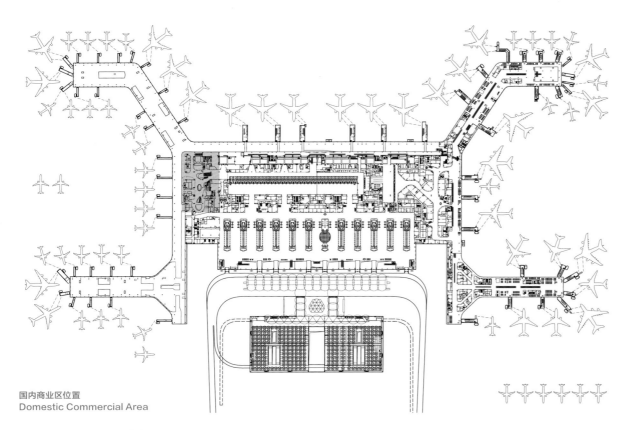

国内商业区位置
Domestic Commercial Area

商业区 以营造非典型商业综合体为基本设计理念，构建融入流程、功能齐全、氛围浓郁的商业设施体系。

1. **全流程商业的嵌入。**动感及趣味性的商业组织，相互嵌套的商业空间，使商业融入旅客流程。创造出行成为享受的可能性，丰富旅客的出行体验；
2. **商业的无边界扩展。**创造开放式的商业业态，以迎合旅客需求的多元化，使二号航站楼成为商业化的目的地与商业活动频繁的公共场所；
3. **依附于流程的商业属性。**出发大厅设置集中餐饮平台；出发流程及到达流程交汇的重要流程节点处设置集中商业，突出建筑与装饰造型，方便旅客辨认；国内混流区域商业实现出发到达共享。

Commercial Area

Following the basic design idea of creating a non-typical commercial complex, the design establishes a self-contained commercial facility system well incorporated into passenger flows.

1. Full-flow embedding. Dynamic and interesting organization well incorporates nested commercial spaces into passenger flows, creating enjoyable trips with diversified experiences;
2. Borderless expansion. Open commercial trades are created to meet passengers' diversified demands and make T2 a commercial destination and a public place with frequent commercial activities;
3. Flow-oriented layout. The design provides centralized F&B platform in the departure hall, centralized commercial zones with eye-catching architectural and decorative shapes at important nodes where the departure and the arrival flows meet, and commercial facilities serving both departure and arrival passengers in the domestic mixed flow area.

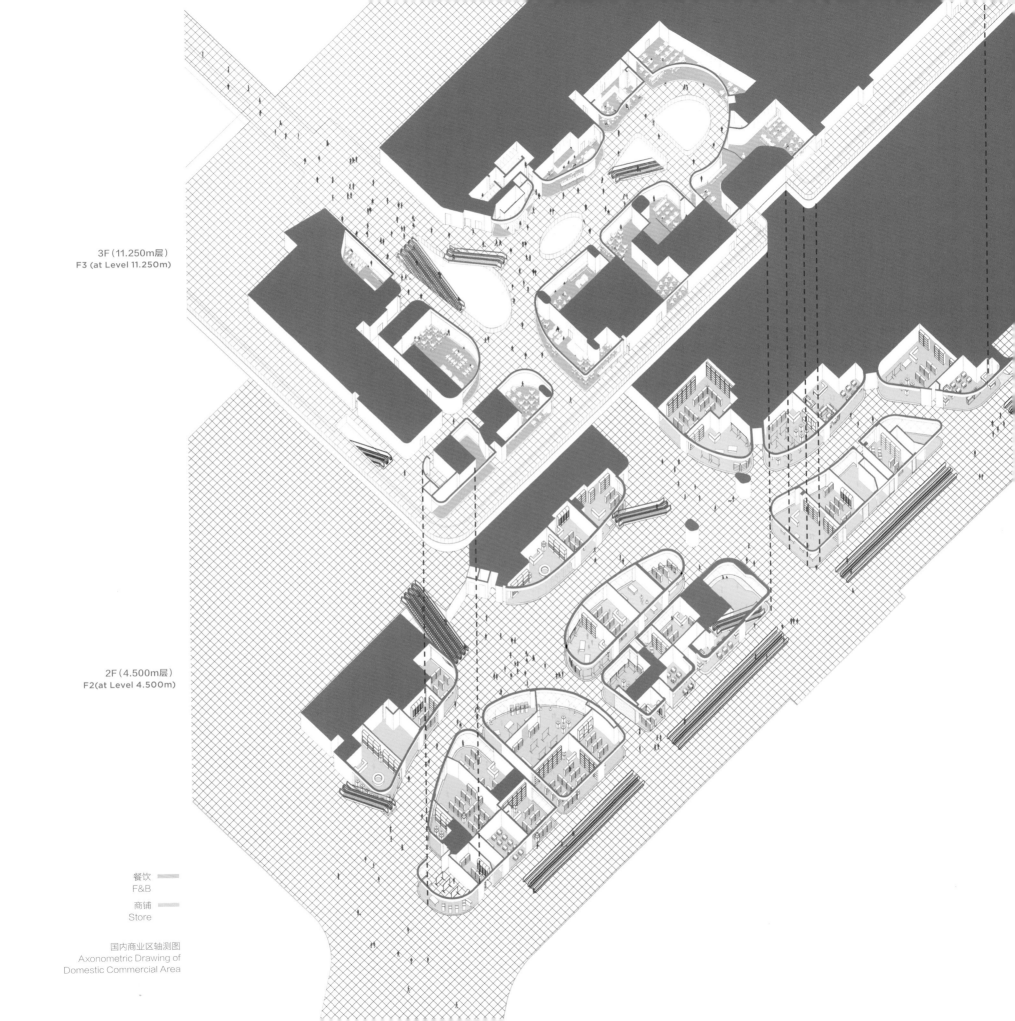

3F（11.250m层）
F3 (at Level 11.250m)

2F（4.500m层）
F2 (at Level 4.500m)

餐饮
F&B

商铺
Store

国内商业区轴测图
Axonometric Drawing of
Domestic Commercial Area

国际商业区
International Commercial Area

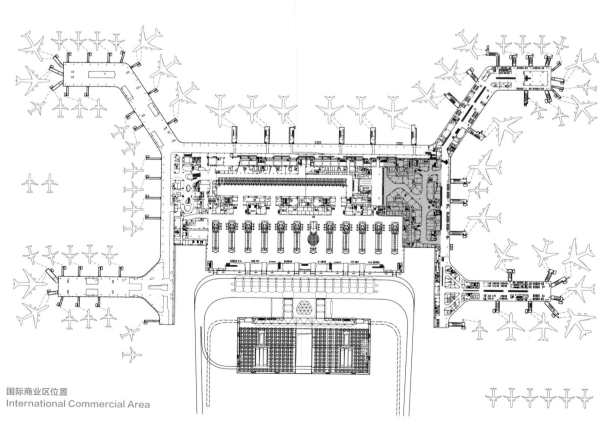

国际商业区位置
International Commercial Area

2F（12.250m层）
F2(at Level 12.250m)

餐饮
F&B

商铺
Store

国际商业区轴测图
Axonometric Drawing of
International Commercial Area

西六指廊国内混流候机厅
West Pier Waiting Hall

东五指廊国际出发候机厅
East Pier Waiting Hall

北指廊国内混流候机厅
North Pier Waiting Hall

国际远机位候机厅
International Remote Stand Waiting Hall

行李提取厅
Baggage Claim Hall

行李提取厅
Baggage Claim Hall

迎客大厅
Arrival Hall

交通中心旅客大厅
GTC Passengers Hall

6.5 大空间照明设计

6.5 LARGE-SPACE LIGHTING

大空间照明根据建筑空间与装修天花特点,采用直接照明为主间接照明为辅的照明方式。人工照明采用LED投光灯、LED筒灯、LED条形灯、LED射灯为主要光源,并将灯具建材化、建筑化实现功能照明,满足规范对照度、功率密度、眩光、显色性等照明指标要求;同时采用少量的间接照明强化建筑的空间感、烘托空间氛围,对一些低矮空间起到消除压抑感的作用。其中,办票大厅还首次尝试使用WRGB可变色的LED灯及间接照明与泛光照明相结合的方式,创造节日模式,为特定日子或有特殊活动时营造更加多样的空间氛围。

6.5.1 设计理念

1. 充分考虑节能、环保,严格控制照明功率密度值,注重照明功能、视觉效果与节能的统一。
2. 根据不同区域空间高度及使用特点采用相应照明及控制方式。
3. 选择高效、节能、环保型灯具、光源及电器附件。
4. 基础照明、应急照明、局部照明、重点照明、广告照明相结合,采用室内功能照明为主、泛光照明为辅,直接照明为主、间接照明为辅。
5. 利用采光天窗与幕墙,将充足的自然采光引入室内,尽量减少白天开灯时间。

6.5.2 办票大厅照明方案

二号航站楼主楼三层办票大厅高大空间平面尺寸约为432m×152m,地面距离天花底部高度最高点约为26m;结构柱为钢柱,柱距为东西向36m,南北向45m,南北两面为玻璃幕墙。结合办票大厅结构、天花形态及空间效果,采用直接照明为主、间接照明为辅的照明方案:

1. 直接照明。结合办票岛的布置将LED深筒射灯藏在天花波浪造型的波峰、波谷吊顶内部,灯具与吊顶天花组合为一体,并将吊顶龙骨、结构桁架与用于灯具安装、检修维护的马道充分融合。灯具安装根据吊顶铝合金板旋转角度调整,通过铝合金板上的圆形透光孔将灯光均匀投射出来,提供大厅功能照明,实现见光不见灯的照明效果。
2. 间接照明。每个办票岛顶部藏着WRGB可变色的LED投光灯,向上打亮作为间接照明与泛光照明,向上照亮顶棚,表现天花曲面造型,并通过铝合金板的漫反射将空间打亮波浪造型的天花,使得照明与建筑和室内设计融合为一体。同时,通过对WRGB灯具的调色控制满足不同的节日气氛需求,为旅客提供不一样的旅行感受。
3. 重点照明。在每个服务柜台上方钢架上局部安装LED筒灯满足办票岛柜台的工作需求。

Based on the characteristics of building space and ceiling finish, direct lighting supplemented by indirect lighting are adopted in big spaces. Main sources of artificial lighting include LED project lamp, LED down lamp, LED strip lamp and LED spot lamp. The lamps are introduced as building materials for functional lighting in accordance with the lighting indicator requirements of relevant codes on illuminance, power density, glare, color rendering, etc. In addition, a small amount of indirect lighting is adopted to strengthen the sense of space and the space atmosphere of the building, and eliminate the depression sense of some low spaces. Among them, the WRGB color-changing LED lamps used in the check-in hall is the first attempt to combine indirect lighting and flood lighting. That creates a holiday mode that can foster more diverse space atmosphere for specific days and special events.

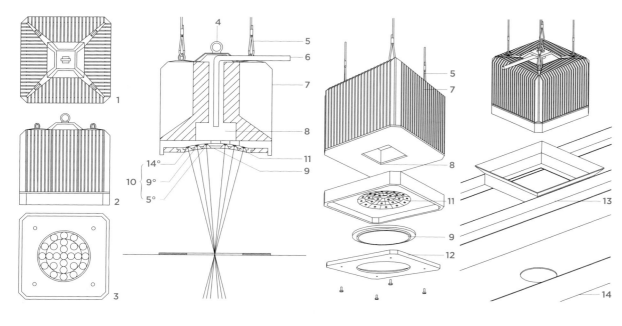

灯具示意图
Schematic Diagram of Lamps

灯具结构剖面图
Structural Section

灯具爆炸图
Blow-up Plan

灯具安装图
Installation Drawing

1. 顶视图
2. 侧视图
3. 底视图
4. 灯具安装吊环
5. 灯具安全绳
6. 电线
7. 散热片
8. 电器位置
9. 透视镜
10. 光源及透视倾斜角度
11. LED
12. 固定透镜零件
13. 灯具安装定位槽
14. 天花

1. Top View
2. Side View
3. Bottom view
4. Lifting Eye for Lamp Installation
5. Lamp Safety Rope
6. Electric Wire
7. Radiator
8. Location of Electrical Appliance
9. Lens
10. Light Source and Oblique Perspective Angle
11. LED
12. Fixed Lens Part
13. Locating Slot for Lamp Installation
14. Ceiling

6.5.1 Design Concept

1. Give full consideration to energy conservation and environmental protection, strictly control lighting power density, and attach importance to the harmony among lighting function, visual effect and energy conservation.
2. Adopt corresponding lighting and control methods according to the spatial height and service characteristics of different areas.
3. Select high-efficiency, energy-saving, environmentally-friendly lamps, light sources and electrical accessories.
4. Reasonably combine basic lighting, emergency lighting, local lighting, accent lighting and advertising lighting; use indoor functional lighting supplemented by flood lighting, and direct lighting supplemented by indirect lighting.
5. Introduce skylight and curtain wall to maximize daylight and minimize daytime artificial lighting.

6.5.2 Lighting in Check-in Hall

The check-in hall is located on F3 of the main building of T2. The planar size of its lofty space is about 432m X 152m; and the height from floor to the highest point of ceiling bottom is about 26m. The structure columns are made of steel in an interval of 36m in east-west direction and 45m in north-south direction. Glass curtain walls are adopted respectively on the north and south facades. Considering the hall structure, ceiling shape and space effect, the design adopts direct lighting supplemented by indirect lighting:

1. Direct lighting. Based on the layout of the check-in island, deep-tube LED spot lamps are concealed in the peaks and troughs of the wave-like suspended ceiling to form an integral whole with the suspended ceiling. Ceiling joists and structural trusses are also fully integrated with the catwalk for lamp installation, maintenance and repair. The installation of lamps is subject to the rotation angle of the aluminum alloy plates of the ceiling, on which round light holes project the light evenly for functional lighting in the hall without exposing the lamps.
2. Indirect lighting. WRGB color-changing LED project lamps are hidden over check-in islands. They are used as indirect lighting and flood lighting to illuminate upward the ceiling to show the curved surface of the ceiling and, through the diffuse reflection of the aluminum alloy plates, brighten the space against the wave-like ceiling, integrating lighting with the building and the interior design. In addition, the lighting color of the WRGB lamps can be adjusted to foster different festival atmosphere, providing passengers with unique travel experience.
3. Accent lighting. LED down lamps are installed in some positions of the steel frame above each service counter to meet the work demands of the check-in island.

办票大厅照明
Lighting in Check-in Hall

办票大厅照明设计概念剖面图
Conceptual Section of Lighting Design for Check-in Hall

安检大厅照明
Lighting in Security Check Hall

边检大厅照明
Lighting in Frontier Inspection Hall

西指廊国内混流大厅照明
Lighting in West Pier Domestic Mixed-flow Hall

东指廊国际出发候机厅照明
Lighting in East Pier International Departure Waiting Hall

6.6 色彩体系设计

6.6.1 整体风格组织

二号航站楼整体色调简洁明快。地面采用浅灰色花岗岩，天花与墙面采用银灰白色铝板，大屋顶天花龙骨喷涂同网架一样的颜色，整体统一；低矮空间镂空天花内部管线、龙骨及结构喷涂深灰色，避免旅客直接看到天花内部。办票大厅、迎客大厅、安检及联检大厅、候机指廊等主要出发空间设置天窗，自然光线从天窗引入，透过天花叶片洒落下来，不同天气形成不同风格的柔和的漫射光充满机场内部；配以外墙大面积玻璃幕墙，与岭南花园等绿化庭院相互渗透；室内配超白玻璃栏杆与隔断，通透自然。

6.6.2 核心空间色彩建构

1. 出发厅北侧玻璃幕墙颜色。分区域设置了红、黄、绿三种颜色的像素画组合渐变玻璃，与花园的绿色交相辉映，犹如透过满洲窗望向幽静的绿化庭院，充满岭南地域特征。
2. 岭南花园。地域自然风格浓郁，营造具有岭南特色、灵动惬意的绿色空间。
3. 安检通道口、行李提取厅口及国际到达联检厅入口绿化墙。绿化墙的设计，给简洁明快的室内带来清新自然的风格，成为空间视觉焦点。
4. 交通节点铺地。在多处核心通道节点及商业区核心区域，设计了黑、红、灰三色组合石材地面，通过各颜色与图案与渐变，建构地面趣味色彩，引导不同功能或方向的空间自然过渡。
5. 交通中心停车楼分区颜色。为提高停车楼辨识度，将交通中心停车楼根据楼层与分区划分为12个区域，每个区域的墙面、柱身及标识均设置不同的颜色加以区分。

6.6.3 系统颜色搭配

1. 大空间照明。办票大厅照明设计可根据时间和人流，针对不同的环境实现不同颜色及照度的照明方案，设有平时和节日两种灯光模式，达到智慧节能与舒适的照明效果。
2. 标识系统颜色。板面底色为深灰色，支撑结构为不锈钢原色。考虑用颜色对流程进行分流指引，对旅客起到暗示作用，不同信息的字体和图标采用不同颜色。具体色彩为：出发、到达作为主流程，图标信息采用黄色；中转信息较为重要，图标信息采用玫红色（同南航中转色）；中英文及小语种文字采用白色；为避免白色信息在淡色装饰面板上或灰白空间内无法凸显，如入口号、办票岛号等数字、字母，采用黄色。
3. 柜台颜色。外表面采用木纹色拉丝不锈钢面板，黑色人造石台面，方便旅客辨识。
4. 座椅颜色。座椅按功能区域选择了5种色系。如主楼三层办票大厅和一层迎客大厅座椅选用明亮的红色系及黄色系，突出色彩点缀效果，寓意对旅客的欢迎；指廊候机区选用浅绿色系，营造温馨轻松的候机氛围。
5. 商业灯箱底色。商业灯箱底色采用深灰色，搭配不同LOGO，突出商业区整体性。

整体浅色明快色调
Bright and Light Color Overall

6.6 COLOR SYSTEM

6.6.1 Overall Style

The color tone of T2 is generally simple and bright. The design uses light grey granite on floors, silver grey-white aluminum plates on ceilings and walls, and the same color of painting for the big roof ceiling joists as the grid, achieving a unified look on the whole. In low spaces, pipelines, joists and structures inside the hollowed-out ceiling are painted dark grey to hide the interior of the ceiling from passengers. For the main departure spaces including check-in hall, arrival hall, security check and CIQ hall, waiting pier, etc., skylights are provided for natural lighting: the daylight sprinkling down through ceiling lamellas can create soft diffused light of different styles under different weather conditions. Plus the extensive glass curtain wall on the façade, the indoor spaces of T2 are perfectly integrated with the outdoor green courtyards including Lingnan gardens. The interior is provided with ultra-white glass handrails and partitions, transparent and natural.

6.6.2 Color Mix for Core Spaces

1. Glass curtain wall on the north façade of the departure hall. Pixel art-painted glass in gradual change of color between red, yellow and green is provided by zones on the north façade. It offers a view into the tranquil green courtyard fraught with the regional characteristics of Lingnan through Manchuria windows.
2. Lingnan gardens. These green spaces are typical of regional natural style of Lingnan, vivid and agreeable.
3. Green walls at the accesses to security check channel, baggage claim hall and international arrival CIQ hall. The green walls bring a fresh and natural style to the simple and bright interior, creating visual focuses indoors.
4. Paving of transportation nodes. The floors of core channel nodes and core commercial zones are paved with stone materials in black, red and grey colors combined. The colors and patterns gradually change, rendering fun mix of colors on the floor and facilitating the natural transition between spaces with different functions or in different directions.
5. Color by zones for the parking building of the GTC. For higher recognizability, the parking building is divided into 12 zones based on floors and functional zones, with different colors for the walls, columns and signage of each zone to distinguish them.

6.6.3 Color Mix for Systems

1. Large space lighting. The lighting system in the check-in hall can realize different mixes of colors and illumination for different environments based on time and people flow. It allows two different modes, i.e. daily use and festivals, and can guarantee both smart energy conservation and comfortable lighting effect.
2. Signage system. The background color of signage boards is dark grey, and the color of supporting structures is the primary color of stainless steel. Colors are used to guide people flows by giving hints to passengers. Specifically, fonts and icons containing different information are colored differently. For instance, icons for departure and arrival, which are main flows, are in yellow; icons indicating transfer information, which is important, are in rose red (same as that for transfer of CSZ); Chinese, English and other minority languages are written in white; figures and letters used to indicate entrance number, check-in island number, etc. on light-colored decorative panels or in grey-white spaces are in yellow instead of white to highlight the information.
3. Counters. The outer surface is made of brushed stainless steel panel in wood grain color and the countertop is made of black artificial stone, convenient for passengers to identify.
4. Seats. Five seat colors are selected as per functional zones. For example, the seats in the check-in hall on F3 and the arrival hall on F1 of the main building are in bright red and yellow with highlighted interspersed effect to greet passengers; the seats in the waiting area of piers are in light green to foster a warm and relaxing waiting atmosphere.
5. Commercial light boxes. Dark grey is used as the background color of commercial boxes in combination with different logos to highlight the integrity of the commercial area.

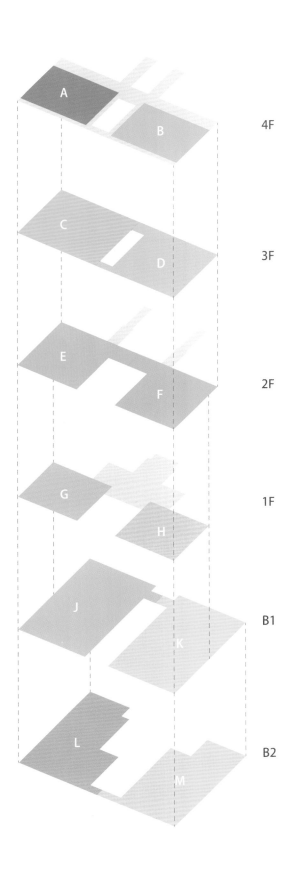

停车楼分区颜色
Zone-based Colors in Parking Building

岭南花园颜色1
Color 1 of Lingnan Gardens

岭南花园颜色2
Color 2 of Lingnan Gardens

室内墙面绿化颜色
Color of Interior Greening Wall

6.7 标识系统设计

6.7 SIGNAGE SYSTEM

办票大厅悬挂标识
Hanging Signage in Check-in Hall

6.7.1 设计概述与目标

在大型公共服务建筑中,特别是交通枢纽型建筑,人们往往因超大空间容易迷失方位。如何让人们快速获取方向信息以及理解色彩、文字、图形的表达,是本工程标识系统设计的核心内容。
1. 以"系统、科学、美观、有效"作为标识系统的设计原则。
2. 以"SKYTRAX"的评定标准为设计标准和依据。
3. 以"一切从旅客感受出发"为设计目标。

6.7.1 General

In large public service buildings, especially those functioning as transportation hubs with extra large space, people tend to get lost. Given this, the key to signage system design for the project is to help people quickly obtain direction information and understand the meanings of color, texts and graphics.
1. Design principle for signage system: systematic, scientific, artistic, and effective.
2. Design criterion and basis: evaluation standard of SKYTRAX.
3. Design objective: all for passengers' experience.

办票大厅立柱标识
Column-mounted Signage in Check-in Hall

6.7.2 设计内容

6.7.2.1 建筑与流程特点分析

1. 建筑特点

（1）出发厅。出发厅空间尺度较大，天花距离地面平均高度为32m，航站楼入口至办票岛的距离达36m，给人宽广的视觉感受。如何将标识形式、标识牌尺度与整体空间做到协调美观是标识设计的首要难题。

（2）候机区。旅客过了联检区（安检、边防、海关）后经过琳琅满目的商业区域，到达候机区。候机区呈线性空间，登机口有序分布在空间外侧，标识指引需要迎合空间形态与信息需求进行序列化设计。突出重点引导、保持信息延续。

（3）迎客大厅。国内、国际迎客大厅空间连通，同时被通往交通中心的二层天桥分隔，导致旅客的部分视线受阻。因此，标识需结合空间形态及视距重点考虑连续引导的问题。

（4）时空隧道。二号航站楼在长达100m的到达中转通道，设置了多媒体光影秀。这在国内机场中，是首创的空间。但从流程功能而言，到达、中转是航站楼内极其重要的通道，标识清晰度及识别度就必须完全高于媒体展示面，才能满足功能引导需求。

（5）交通中心停车楼。停车容量非常巨大，总层数达6层，每层停车数量近千个。除了引导车辆进出，旅客停车后寻找航站楼、到达旅客反向寻车都是标识设计的重点。同时，多层停车楼必须对其分区分色，才能使千篇一律的停车空间给人留下印象。

2. 旅客流线

（1）禁区外单向流线。虽旅客从航站楼入口由南向北进行办理乘机手续、安检一系列登机前的过程，流程中途经商业、洗手间、服务柜台设施。这样的单向流线为标识逐级引导带来了便利。

（2）禁区内出发、到达、中转混流。二号航站楼国内为混流流程，对于标识引导来说是一个挑战。国内区域的标识均为双向标识，既有出发登机口指引，又有到达中转指引，同时为了中转流程更为有效快捷，考虑从颜色上进行规划处理。

（3）远距离移动。旅客进入候机区域，商业与服务的体验可能造成旅客远距离移动，因此候机区域可能并不是传统的等候空间，我们将它看作是一个活动空间。因此，标识的引导必须满足商业空间行人的特点，即随处可达想去的地方。

6.7.2 Design Contents

6.7.2.1 Architectural and Flow Characteristics

1. Architectural Characteristics

(1) Departure hall. The departure hall has a large space: the average height from floor to ceiling is 32 m and the distance from the entrance of the terminal to the check-in island is 36 m, spacious visually. The primary difficulty in signage design is to harmonize signage form and signboard scale with the overall space.

(2) Waiting area. After passing through the CIQ area (for security check, frontier inspection and customs check) and then the dazzling commercial area, passengers arrive at the waiting area. The waiting area is a linear space, with boarding gates orderly distributed on the outside, so the guide signs should be serialized to meet the spatial form and information demands to highlight the key orientation information and maintain information continuity.

(3) Arrival hall. The domestic and international arrival halls are connected to each other but separated by the double-tiered overpass leading to the GTC, blocking some sight of passengers. Therefore, it is necessary to ensure continuous orientation in consideration of the spatial form and sight distance.

(4) Time tunnel. T2 provides a multimedia light show in the 100m-long arrival and transfer channel, which is the first of its kind in domestic airports. In terms of flows, arrival and transfer channel is the most important channel in the terminal, so the degree of signage legibility and identification must be higher than that of media display to support functional orientation.

(5) Parking building of GTC. The parking building has a huge parking capacity with nearly 1,000 parking spaces on each of the 6 floors. In addition to guiding vehicles in and out, the focus of signage design also includes departure passengers' looking for the terminal after parking, and the arrival passengers' looking for parked vehicles after arrival. Besides, the multi-story parking building must be zoned by different colors for the parking spaces to impress passengers.

2. Passenger Flows

(1) One-way flows outside restricted area. From the entrance of the terminal on the south to the boarding gate on the north, passengers go through a series of pre-boarding procedures including check in and security check while passing by commercial zones, washrooms, service counters and the like. Such one-way flows enable easy step-by-step orientation of the signage system.

(2) Mixed departure, arrival and transfer flows inside restricted area. The domestic area of T2 adopts mixed flows, bringing a challenge to signage orientation. Signs in domestic area are all two-way ones offering orientation for both departure boarding gates and arrival/transfer flows. The signage system is designed in different colors to allow for more efficient and faster transfer.

(3) Long-distance movement. After entering the waiting area, the commercial and service facilities might attract passengers to move a long distance. In view of that, the waiting area is deemed as a dynamic space rather than mere waiting. Therefore, the signage system must be able to guide commercial pedestrians to anywhere they want.

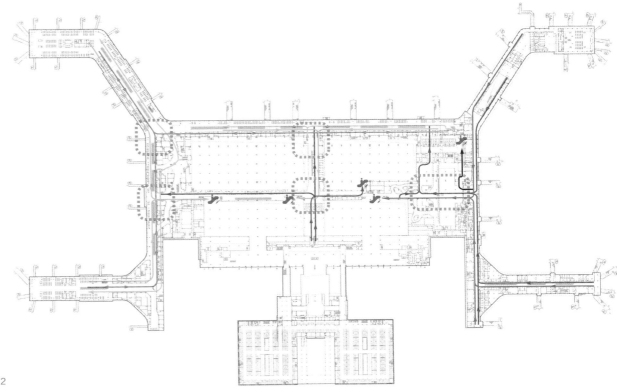

二层混流区域流线及标识节点示意
Diagram of Flow and Signage Nodes in Mixed-flow Area of F2

国内出发
Domestic Departure

国内到达
Domestic Arrival

国际出发
International Departure

国际到达
International Arrival

中转- 国内转国内
Domestic-Domestic Transfer

中转- 国内转国际
Domestic-International Transfer

中转- 国际转国内
International-Domestic Transfer

中转- 国际转国际
International-International Transfer

关键节点
Key Node

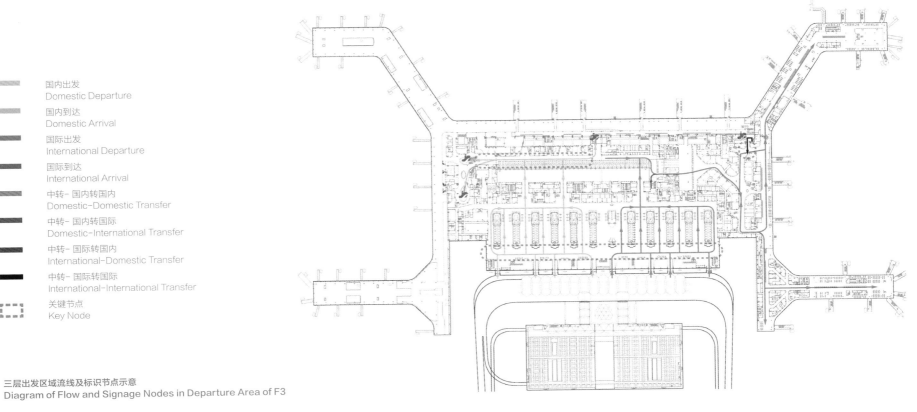

三层出发区域流线及标识节点示意
Diagram of Flow and Signage Nodes in Departure Area of F3

6.7.2.2 信息范畴与标识类型

1. 信息设计元素

标识牌上的信息由图形、文字组成，图形包含箭头、特定图标、说明性图示等，文字包含中文、外文。这些信息的组合向旅客传递流程引导、设施服务、说明告示、警告禁令等内容的导向管理。

（1）字体。黑体作为大型公共建筑的常用字体显得富有现代感，它的等线、棱角等笔画特点在识别度上占很大优势。同时，我们也模拟了发光效果、旅客视力效果，作为选择的力证。最终我们将微软雅黑作为中文字体，Frutiger作为英文字体。

（2）语言。按照实现"SKYTRAX五星航站楼"的认证目标，除中英文外，标识指引增加了小语种，包括法文、日文、韩文。这些语言设置的原则为：小语种仅出现在国际流程的方向指引类标识中。考虑信息表达的简洁性和面板尺寸的局限性，"登机口"作为国际上常见信息，仅显示中英文，不显示小语种。

2. 标识类型

根据引导标识在航站楼中所起作用，大致可分为两大类，一类用于固定流程引导和目的地指示的静态标识，另一类是用于航班显示的动态显示屏。

6.7.2.3 标识布点原则

1. 航站楼

（1）标识点位必须在旅客主要动线的交汇处；
（2）在平面流线和垂直上下流线的交汇处需要设置标识牌；
（3）标牌设置尽可能垂直于旅客视线方向；
（4）在满足引导功能的前提下，尽可能合理控制布点数量；
（5）点位设置需同时考虑信息内容的连贯，做到信息链完整；
（6）标识点位必须考虑建筑内的服务功能设施，与环境协调统一。

2. 交通中心及停车楼

交通中心及停车楼，旅客可以在这里轻松实现飞机、地铁、城轨、大巴、出租车、私家车等多种交通方式"零距离换乘"。标识布点采用的原则为"由近至远、逐项分流"，在信息指引的方式上也根据此原则排列信息。

6.7.7.2 Information Category and Signage Type

1. Information Design Elements

The information on the signboards consists of graphics and texts, of which the graphics contain arrows, specific icons, explanatory chart, etc. and the texts are in Chinese and foreign languages. The combination of all these conveys such orientation information as flow guidance, facility service, explanatory notice, warning and ban, etc. to passengers.

(1) Font. As a font commonly used in large public buildings, boldface appears modern and, due to its equal line, corner angle and other stroke characteristics, advantageous in identification. With simulations of luminous effect and passenger's visual effect as strong proof, Microsoft Yahei is determined as Chinese font and Frutiger as English font.

(2) Language. To pursue the certification of "5-star terminal rated by SKYTRAX", in addition to Chinese and English, minority languages including French, Japanese and Korean are added to the signage system. The principle for language setting is: minority languages only appear in the direction-indicating signs of international flows. In view of simplicity of information expression and limited panel size, "boarding gate" as commonly seen information is indicated only in Chinese and English.

2. Signage Type

There are generally two types of orientation signage defined by their roles in the terminal: static signage used for fixed flow orientation and destination indication, and dynamic displays to show the flight information.

6.7.2.3 Signage Distribution Principles

1. Terminal

(1) The signs should be placed at where the main passenger circulations meet;
(2) The signs should be placed at where the planar and vertical circulations meet;
(3) The signs should be placed perpendicular to passengers' sight line where possible;
(4) The number of the signs should be controlled as reasonably as possible on the condition of satisfying the orientation function;
(5) The coherence of information content should be duly considered in signage distribution to maintain the continuity of information chain.
(6) The service and functional facilities inside the building should be duly considered in signage distribution to keep them harmonious with the environment.

2. GTC and Parking Building

In the GTC and parking building, passengers can enjoy "zero-distance interchange" provided by various transportation means such as airplane, rail transit, bus, taxi, private car, etc. Here, the principle for signage distribution and guidance information arrangement is "from near to far, flow diversion by item".

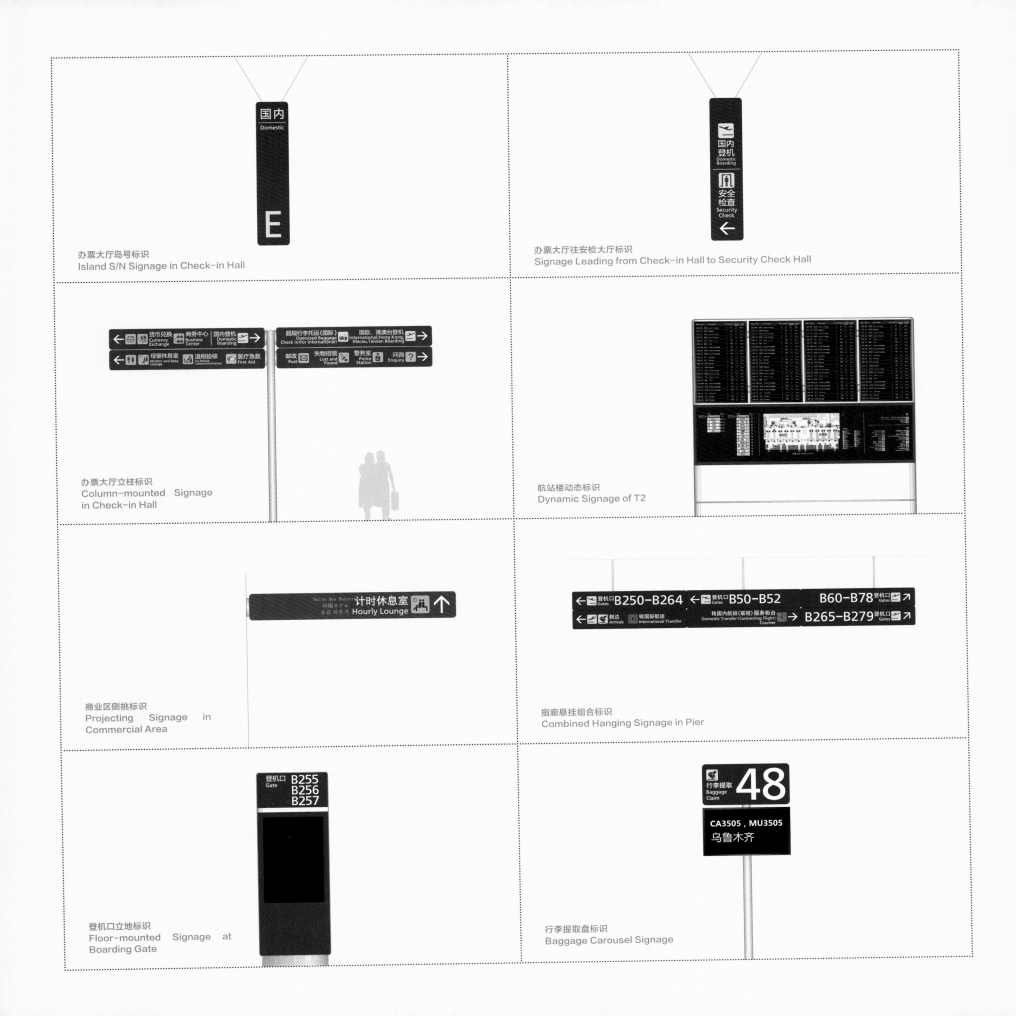

6.7.3 主要设计手法

6.7.3.1 色彩区分流程
考虑用颜色对流程进行分流指引，通过多颜色搭配比选，确定色彩方案：
1. 出发、到达作为主流程，图标信息采用黄色；
2. 中转信息较为重要，图标信息采用玫红色（同南航中转色）；
3. 中英文及小语种文字采用白色；
4. 为避免白色信息在淡色装饰面板上或灰白空间内无法凸显，如入口号、办票岛号等数字、字母，采用黄色。

6.7.3.2 大空间大标识
在整个航站楼标识系统中，有三处大空间大标识的处理方式：
1. 办票大厅办票区标识；
2. 办票后方安检及登机口指引；
3. 国内混流区指廊交接点指引。

6.7.3.3 人性化设计（步行提示）
在建筑分析中提到远距离移动的特点，这样的移动路径容易给旅客造成心理上的不安和慌乱。为此，我们在设计中加入了步行距离提醒，为候机区域的远距离移动提供了提前告知的义务。同时，在旅客流程的重要判断节点上，采用了直观的步行时间提醒。

6.7.3.4 标准化设计（洗手间标识、登机口标识）
1. 洗手间。二号航站楼内每个洗手间配有标准的标识布置方式，入口门头采用侧挑式三角标识，为旅客提供远距离识别的作用。进入洗手间通道，男女各方向上的通道上采用较大的人形图案标识，通道尽头均设置小型门牌加以重复告知。母婴室及无障碍洗手间在门旁一侧设置门牌标识；
2. 登机口。将登机口标识看作是一个小区域模块，同时配有座位区登机口号标识、登机口柜台航显标识，无论柜台与座椅如何摆放，登机口号标识、登机口柜台航显标识都会组合出现。

6.7.3.5 模块化设计（航显、灯箱）
延续建筑的模数化设计，规整标识体型。特别在动态航显标识的样式中，用模块化的方式排列屏幕，既美观整齐又维护便利。同样，由于航站楼信息指引内容较多，单个灯箱并不能满足信息排列要求，因此模块化的灯箱组合是最佳的展现信息的方式。

6.7.4 编号命名设计

6.7.4.1 出入口
一号航站楼的出入口编号为1~36，由于二号航站楼与一号航站楼建筑主体相连接，二号航站楼的出入口与一号航站楼不重复。考虑未来一号航站楼的可变因素，预留37~40，二号航站楼从41起编，与一号航站楼原则由西向东、由三层至一层排序编号相同。

6.7.4.2 办票区/柜台
办票区编号由于二号航站楼办票区对于旅客为竖向分布，办票岛的划分方式以通道为原则划分，既能减少旅客寻找柜台的时间，又能将大厅两例的贵宾办票柜台与普通旅客柜台区分开来。
编号从西向东排序编号，留出字母A、B给登机口，从C起编号（跳开字母I和O），即两侧贵宾办票为C和Q普通旅客办票为D~P。
对于运营部门提出的管理及分配办票柜台问题，内部可按"岛"进行管理，如"1号岛"、"2号岛"等。
每个办票区域的柜台编号，从01依次起编。将开包间纳入编号体系，方便人员口头招引。

6.7.4.3 登机口
登机口编号（近机位）
二号航站楼登机口编号延续一号航站楼站接登机口的原则：
1. 与机位号一致；
2. 延续一号航站楼航站楼登机口AB原则；
3. 登机口（国际）为A区，西边登机口（国内）为B区。
登机口编号（远机位）
远机位预留号码给每条指廊作远期规划考虑。

6.7.4.4 行李提取转盘及到达出口
一号航站楼的行李转盘号为1~22，为确保两个航站楼内编号的唯一性，同时考虑未来一号航站楼的可变因素，二号航站楼行李转盘从西向东，由31起编，国内行李转盘号31~41，国际行李转盘号42~51。

6.7.4.5 停车楼
1. 一号航站楼周边的停车场有P1（东停车场）、P2（南停车场）、P3（北停车楼）、P4（北停车场）；
2. 二号航站楼周边新建停车场和停车楼的编号，从P6起编；
3. 私家车室外停车场命名为P6（东）、P7（西）；
4. 停车楼由两栋楼组成，但出入口为东楼进、西楼出，因此停车楼统一编号为P8；
5. 区域编号。由于P8停车楼被通道划分为两个区域，因此每层用两个字母区分，从上至下顺序编号。为确保旅客所在空间的视觉唯一性，每个字母对应一个颜色，帮助旅客识别和记忆，同时增添停车楼的明亮度。为避免颜色相近造成误解，配以图案作为记忆辅助。

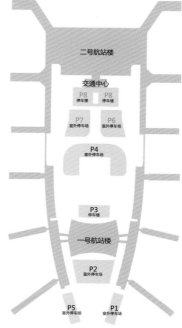

航站楼出入口编号图
Terminal Accesses Numbering Diagram

停车场编号图
Parking Lot Numbering Diagram

6.7.3 Main Design Approaches

6.7.3.1 Color-defined Flows
The color scheme based on multi-color matching and comparison guides different flows using different colors:
1. Icons for departure and arrival, which are main flows, are in yellow;
2. Icons indicating transfer information, which is important, are in rose red (same as that for transfer of CSZ);
3. Chinese, English and other minority languages are written in white;
4. Figures and letters used to indicate entrance number, check-in island number, etc. on light-colored decorative panels or in grey-white spaces are in yellow instead of white to highlight the information.

6.7.3.2 Large Signage for Large Spaces
The signage system of the terminal provides large signs for three large spaces.
1. Signs for the check-in area in the check-in hall;
2. Guide signs for security check and boarding gate after check-in;
3. Guide signs for pier joints in domestic mixed-flow area.

6.7.3.3 Passenger-friendly Design (walking distance reminders)
The building analysis mentions the characteristics of long-distance movement. Such moving paths likely cause psychological anxiety and confusion to passengers. In view of this, walking distance reminders are added in the design to provide early notification of long-distance movement in the waiting area. In addition, directly visualized walking time reminders are provided at important judgment nodes in passenger flows.

6.7.3.4 Standardized Design (washroom and boarding gate signage)
1. Washroom. In T2, each washroom is provided with typical signage layout. This includes a side-mounted triangle sign at the entrance for passengers to identify from afar, a large human figure sign respectively at the access to male and female washroom to point directions and a small sign at the end of the access respectively for repeated notification, and doorplates on the side of the doors of the baby care room and barrier-free washroom;
2. Boarding gate. The boarding gate signs can be regarded as a small-area module, equipped with signs of boarding gate number in seating area and flight display at boarding gate counter. The boarding gate number signs and the flight display signs at the boarding gate counter appear as a combination regardless of the layout of the counter and seats.

6.7.3.5 Modular Design (flight information display and light box)
The signage design continues the modular design of the building to present standardized sizes of signs. The dynamic flight display signs, in particular, are arranged on screen in a modular manner, aesthetic in appearance and convenient for maintenance. Similarly, modular combination of light boxes serves as the best way to present information given that there are so many contents requiring guidance in the terminal and a single light box cannot meet the information arrangement requirements.

6.7.4 Numbering Design

6.7.4.1 Entrances and Exits
For T1, the entrances and exits are numbered 1 to 36, with 37 to 40 reserved for possible change in future. For T2 which is connected with the main building of T1, the entrances and exits are numbered from 41, without repeating the figures used by T1. Following the numbering principle of T1, the numbering in T2 is conducted from west to east and from F3 to F1.

6.7.4.2 Check-in Areas/Counters
For numbering in check-in areas of T2, as the check-in islands are vertically distributed to passengers, they are divided by channel, which can not only reduce the time for passengers to find the counters but also distinguish the VIP check-in counters at both sides of the hall from the ordinary passenger counters.
The counters are numbered from C (skipping letter I and O) from west to east, with letter A and B reserved for boarding gates; i.e., the VIP counters are numbered by C and Q and the ordinary passenger ones by D to P.
As for the check-in counter management and distribution matters concerned by the operation department, the counters can be managed by "islands", e.g. "Island 1", "Island 2", etc. Each check-in area, the counters are numbered from 01 onwards. The unpacking room is incorporated into the numbering system to facility oral guidance.

6.7.4.3 Boarding Gates
Boarding gate numbering (for contact stands)
The numbering of boarding gates of T2 follows the principle of T1:
1. The boarding gate number is consistent with the stand number;
2. The numbering follows the AB principle for the boarding gates of T1;
3. The (international) boarding gates are referred to as Zone A, and the (domestic) boarding gates on the west as Zone B.
Boarding gate numbering (for remote stands)
The remote stand numbers are reserved for each pier for long-term planning.

6.7.4.4 Carousels and Arrival Exits
Because the carousels of T1 are numbered 1 to 22, to ensure the uniqueness of the numbers in the two terminals and leave room for possible change in future in T1, the carousels of T2 are numbered from 31, among which those for domestic flights are numbered 31 to 41 and those for international flights 42 to 51.

6.7.4.5 Parking Building
1. The parking lots around T1 include P1 (east parking lot), P2 (south parking lot), P3 (north parking lot) and P4 (north parking lot);
2. The new parking lots and parking building around T2 are numbered from P6;
3. The outdoor parking lots for private cars are numbered P6 (on the east) and P7 (on the west);
4. The parking building consists of two buildings with the entrance in the east building and exit in the west building. It is numbered P8;
5. Zone numbering. Because P8 (parking building) is divided into two zones by channel, each floor is numbered using two letters and all the zones are numbered sequentially from top to bottom. To ensure the visual uniqueness of the space where passengers are located, each letter corresponds to a color to help passengers identify and remember while enhancing brightness of the parking building. In addition, patterns are used as supplementary approach for position guidance to avoid misunderstanding caused by similar colors.

6.8 广告设计

6.8 ADVERTISEMENT

1. 以服务流程为主的广告点位分布原则。广告点位在设计时考虑结合出发、到达、中转等流程合理分布,依附于流程。媒体的位置、尺寸、亮度等因素均不能影响旅客行程,特别是涉及办票、安检、边检等重要检查区域时不可设置影响工作人员或旅客操作的广告元素。

2. 以标识系统优先的原则。航站楼的室内设计中,视线分析的结果往往会出现广告与标识系统的重合,此类情况应以满足旅客指引的标识系统为先。

3. 与室内其他元素相平衡的原则。航站楼广告点位是质与量的平衡,广告的设置并不是数量越多效果越好,如同构图的留白与法餐的装盘。经过多轮的室内空间模型模拟与实地样板的比对,我们根据室内空间的效果舒适度删除了部分影响室内品质的点位,重点空间或适宜区域也通过运用局部增加广告的尺寸以提高空间的表现张力。

4. 与装修的一体化设计的原则。广告的视觉面是其售出价值的评判标准之一,这也使得广告往往占据了室内空间较大的面积。在进行航站楼室内设计时,我们将广告形式与室内装饰板或造型进行一体化考虑。广告面域的比例尺寸与装饰墙面的构件、模数契合设置,直线与曲面的内墙面均可完美融合媒体屏幕。同时针对广告媒体的构造、材质、颜色、收口节点等外观内在要素,均控制其与相邻装饰构件的协调一致。

1. Flow-oriented distribution. The advertising points should be distributed properly in consideration of the flows of departure, arrival and transfer etc.. The position, size, brightness and other factors of media should not disturb pre-boarding procedures; advertising elements that may hinder operation of airport staff and passengers should be particularly avoided in critical areas such as those for check-in, security check and frontier inspection.

2. Signage system priority. Visual analysis in terminal interior design may frequently indicate the overlapping of advertising system and signage system, in which case signage system should be prioritized to ensure proper orientation for passengers.

3. Balance with other interior elements. Advertising points in the terminal requires both quality and quantity. "The more" is not always "the better", that is why we have blanks in picture composition and French dish plating. Through rounds of interior spatial modeling and simulation as well as on-site sample comparison, some advertising points are removed to improve interior spatial effect and design quality, while some advertising points in important space or proper areas are enlarged to strengthen spatial tension performance.

4. Integrated design with the finish. Visual surface is one of the judging criteria for the selling value of an advertisement. That explains the usually large area occupied by advertisement in interior space. The interior design of the terminal considers the advertising form and interior decorative boards or models as a whole. It aligns the scales of advertising areas with components and modules of decorative walls, so that both straight and curved interior walls are well integrated with media screens. It also coordinate the appearance factors of advertising media such as composition, material, color and close-up node with their neighboring decorative components.

国内商业区广告设计
Advertisement in Domestic Commercial Area

主楼广告设计
Advertising Design in Main Building

指廊广告设计
Advertising Design in Pier

6.9 文化设计

6.9 CULTURAL FEATURES

岭南花园让出发旅客在现代化的航站楼内可以感受到传统园林的魅力,是人与自然融合的最佳实践。身处这里,犹如漫长旅途的一片绿洲,体现了广州地处中国南方的气候特点。岭南花园绿化墙面提取体现岭南特色窗、墙元素,是传统与现代的融合。

"宇宙飞船"——办票大厅的文化广场,彰显独特的公共文化魅力,成为展示公共文化艺术和商业的独特平台。也可以举办商业推广活动。

岭南花园、时尚及传统多维的体验、公共艺术、独特的标志性和文化内涵、流畅的旅客体验、时空隧道使旅客体验时空穿梭的绚丽感觉,共同构建机场文化特色。

Lingnan gardens allow passengers to experience the charm of traditional gardens in a modern terminal setting. This is the best practice to integrate human and nature. Being here is like encountering an oasis typical of the southern China climate of Guangzhou during a long journey. The greening walls of Lingnan gardens are designed with the unique window and wall elements of Lingnan, displaying an integration of tradition and modernity.

The "Spaceship" culture square of the check-in hall serves as a special platform for displaying the unique charm of public culture, art and commerce, and for staging commercial promotions.

The Lingnan gardens, multi-dimensional experience of fashion and tradition, public art, unique representativeness and cultural connotation, smooth passenger experience, and time tunnel for passengers to experience the wonder of time travel, jointly constitute the cultural features of the Airport.

岭南花园
Lingnan Gardens

时空隧道
Time Tunnel

文化广场
Cultural Square

博物馆
Museum

6.10 FUNCTIONAL FACILITIES FOR PASSENGERS

T2 is designed as per the estimated passenger traffic of 45 million and the desired service level of no lower than Class C of IATA.

6.10.1 Tensile Membrane Canopy

Tensile membrane canopy is a unique and representative element of Baiyun Airport in consideration of the hot climate with abundant rainfall of Lingnan. The tensile membrane canopy of T2 covers a floor area of 30,000m² with the largest span of the steel structure being 34 m. It is composed of six areas, including the viaduct of the main building, the top floor curbside of the GTC, the F1 parking lot of the GTC, etc.. The tensile membrane canopy is designed in continuous multi-span arc shape echoing with the continuous arc eave of south facade. The diffuse reflection of the membrane gently reflects the metal texture of the eave, creating a diversely-layered rolling cloud image with the change of light and shadows. The framework+ tensile membrane structure and saddle-shaped hyperbolic-parabolic surfaces comply with mechanical requirements. The designed service life of PTFE resin (PTFE-A) cladding is 30 years (15-year warranty).

The highest point of the tensile membrane canopy of the viaduct of the main building is about 11 m off the viaduct ground. Glass skylights are provided between the spans of the membrane. Water and electricity sleeves are hidden inside the steel column and holes are reserved at the column head of steel casting. The reserved water and electricity pipelines in the column are constructed simultaneously with steel column, whose top section connects to the siphonic roof outlet, lateral side connects to the lights around the column head, and bottom section connects precisely to the concrete foundation, drainage system and illumination system interfaces of the viaduct. The column is poured with asphalt to separate drains from the steel column structure and prevent the latter from internal corrosion.

Departure Curbside Tensile Membrane Canopy

出发车道边张拉膜雨蓬鸟瞰
Bird's Eye View of Departure Curbside Tensile Membrane Canopy

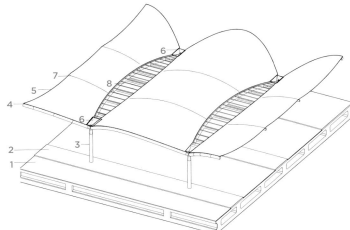

雨蓬标准单元
Typical Canopy Unit

雨蓬与排水、照明组合构造
Composition Construction of Canopy and Drainage & Lighting

1. 高架桥路面
2. 人行通道
3. 雨蓬钢立柱
4. 雨蓬结构主钢管
5. 聚四氟乙烯树脂膜材
6. 不锈钢雨水斗
7. 雨蓬结构次钢管
8. 钢化夹胶彩釉玻璃
9. 照明灯具
10. 雨水管

1. Viaduct Surface
2. Pedestrian Passage
3. Steel Column of Canopy
4. Main Steel Tubes of Canopy Structure
5. PTFE Resin Film
6. Stainless Steel Roof Drain
7. Secondary Steel Tubes of Canopy Structure
8. Laminated and Tempered Fritted Glass
9. Lighting Fixtures
10. Storm Sewer

6.10.2 门斗

门斗采用轻盈的外飘檐造型，与金属屋面及檐口协调一致；外饰面采用白色彩釉玻璃，贴近白云的朦胧感，材质光感与幕墙融为一体；内饰面采用压纹不锈钢，与玻璃形成强烈对比；收边收口力求极致轻薄，展现材料及形体转接的纯粹性。每个门斗对称设置2个自动门通道，在防爆安检常态化情况下，能同时满足防爆安检入口及出口布置要求，互不干扰，减少旅客步行距离。

6.10.2 Foyer

The foyer is designed in the form of light projecting eave, consistent with the metal roof and eave. Its exterior finish adopts white fritted glass that is close to the hazy feel of cloud and features the same brightness as the curtain wall; and its interior finish uses embossed stainless steel in sharp contrast with glass. All its close-ups are extremely light-weighted to demonstrate the purity of material and shape connections. Each foyer is provided with two automatic door accesses symmetrically to, in the context of the normalization of anti-explosion security check, accommodate the layout of both entrance and exit of security check without mutual interference and with minimal walking distance.

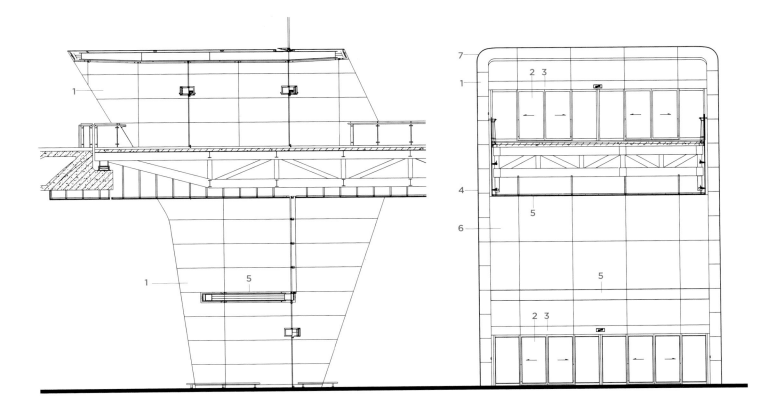

1. 2mm拉丝不锈钢板
2. 电动门
3. 门头灯箱
4. 钢化夹胶彩釉玻璃
5. 3mm厚铝单板
6. 超白钢化夹胶中空玻璃
7. 弯弧夹胶彩釉玻璃

1. 2mm Brushed Stainless Steel
2. Electrically Operated Gate
3. Door Head Light Box
4. Laminated Tempered Fritted Glass
5. 3 mm Thick Aluminum Panel
6. Ultra-Clear Laminated Tempered Insulating Glass
7. Arched Laminated Fritted Glass

门斗内景
Interior View of Foyer

门斗外景
Exterior View of Foyer

6.10.3 办票岛

1. 集成一体化的设计。相比传统的单元化办票岛设计,二号航站楼办票岛采用集成一体化外形,高度整合建筑、结构、机电、弱电、装修、行李与安防等各专业功能;
2. 具有航空器动感外形的设计。结合航空飞行器的特点,采用楔形线条+曲线,营造动感且具冲击力的非线性岛体外形,体现航空运输的速度感;
3. 便于施工的细节设计。三维曲面二维化,使用单曲面取代双曲面,通过组合形成丰富的造型,便于施工。

6.10.3 Check-in Island

1. Integrated design. Compared with traditional unitized check-in island, the check-in island of T2 shows a unified shape, highly integrating architecture, structure, MEP, ELV, finish, baggage, security and other specialized functions;
2. Dynamic aircraft shape. Based on the characteristics of aircrafts, the design uses wedge-like lines and curves to shape up a dynamic non-linear island shape with strong visual impact to reflect the sense of speed of air transportation.
3. Construction-friendly details. The design turns 3D curved surfaces to 2D ones and replaces double curved surfaces with single curved ones, creating diversified shapes through combinations to facilitate construction.

办票岛
Check-in Island

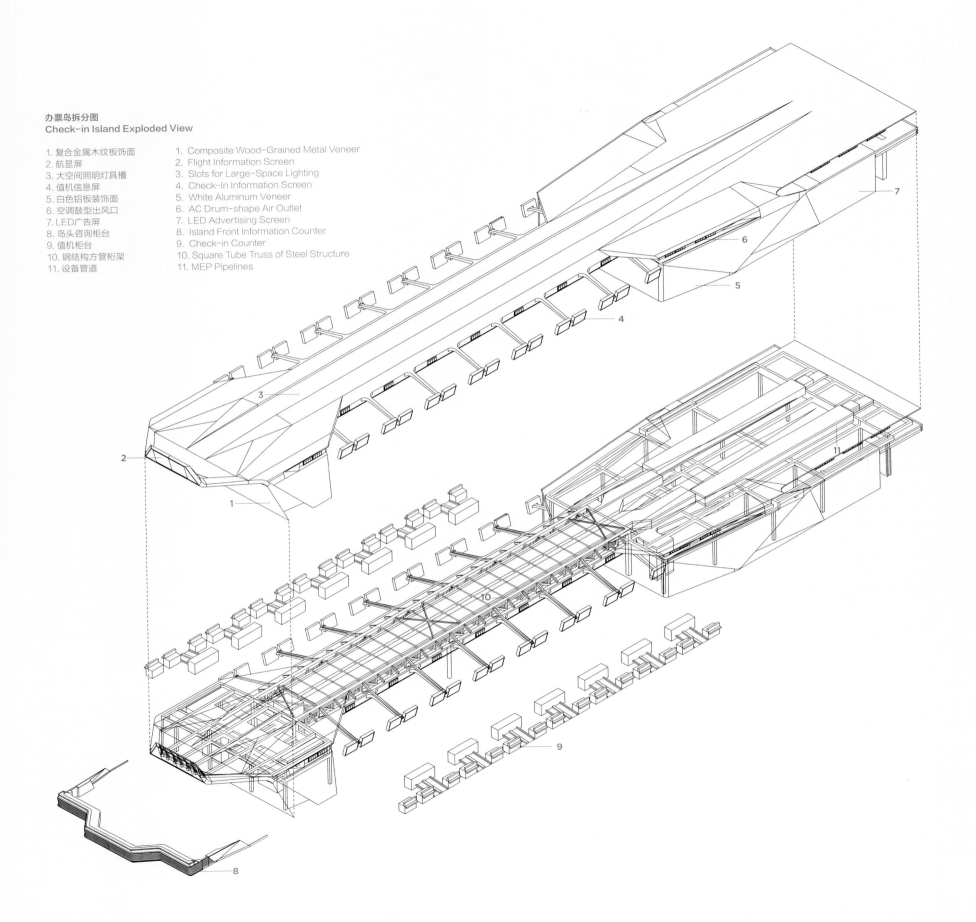

6.10.4 安检通道

安检通道采用玻璃隔断灵活分隔,适应国际国内安检通道资源错峰调配需要。安检通道单元主要由安检设施与隔断设施组成。安检设施包括X光机前后滚轴行李台、验证台、安检摄像、安检屏幕、预留强弱电接口、固定支撑杆件等设施。隔断采用不锈钢立梃及渐变岭南纹饰彩釉玻璃,平衡兼顾安检封闭要求与旅客体验,并体现了岭南地方特色。上述设施结合整体装修风格,进行组合化、标准化整合设计。

6.10.4 Security Check Channel

Security check channels are equipped with glass partitions to flexibly serve peak-time international or domestic security check. A unit of security check channel mainly includes security check and partition facilities. Security check facilities include X-ray baggage inspection bench with front and back rollers, verification counter, security check camera, security check screen, reserved HV and ELV interfaces, fixed supporting rods, etc.. Partitions are made of stainless steel studs and fritted glass with Lingnan-style ornamentation in gradual change, giving equal consideration to enclosure required for security check and passenger experience while reflecting Lingnan characteristics. The aforesaid facilities are standardized and integrated in design in consideration of the overall finish style.

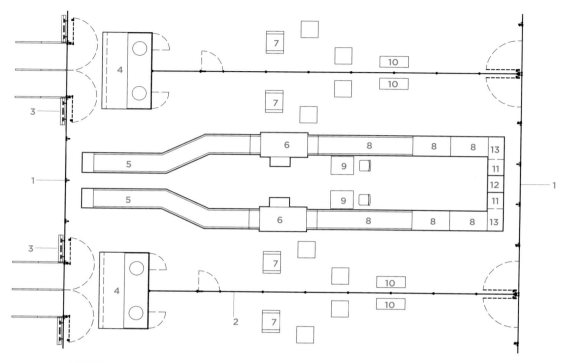

安检通道平面
Plan of Security Check Channel

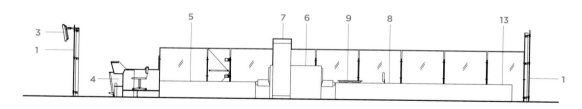

安检通道剖面
Section of Security Check Channel

1. 渐变图案彩釉玻璃　　1. Fritted Glass in Gradation Pattern
2. 乳白色PVB玻璃　　　2. Opalescent PVB Glass
3. 航显屏　　　　　　　3. Flight Information Screen
4. 安检柜台　　　　　　4. Security Check Counter
5. 前传台　　　　　　　5. Conveyor Guide Counter
6. X光机　　　　　　　6. X-ray Machine
7. 安全门　　　　　　　7. Security Door
8. 开包台　　　　　　　8. Unpacking Table
9. 判图台　　　　　　　9. Image Screening Counter
10. 穿鞋凳　　　　　　10. Foot Stool
11. 炸探机柜　　　　　11. Bomb Detector Cabinet
12. 后联络口　　　　　12. Rear Contact Counter
13. 开包工作站机柜　　13. Unpacking Workstation Cabinet

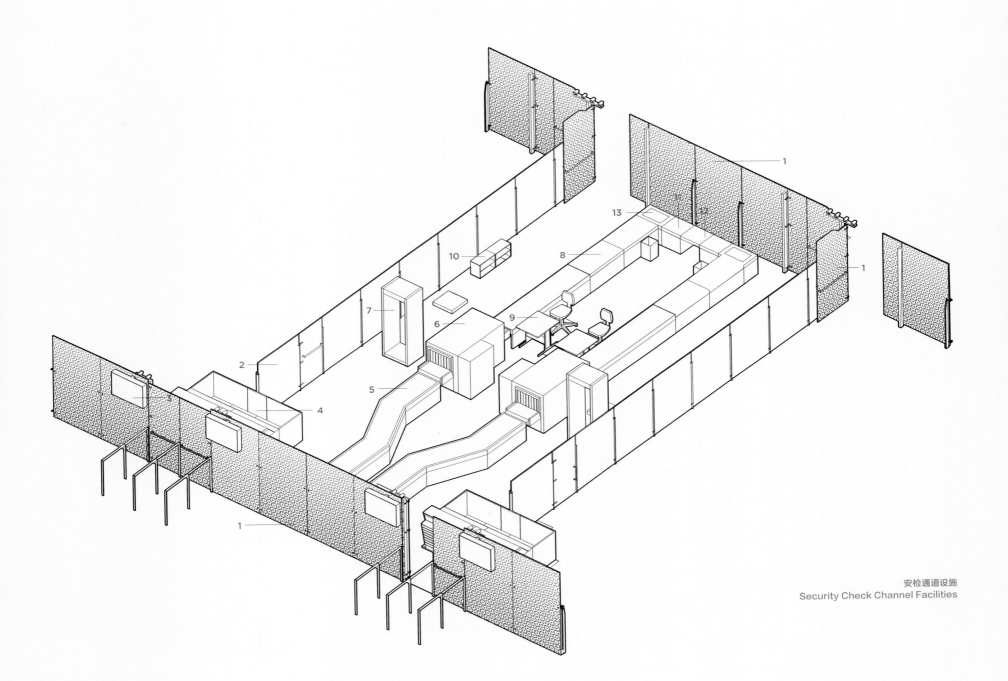

安检通道设施
Security Check Channel Facilities

主楼四层餐饮平台
F4 F&B Platform of Main Building

指廊商业
Pier Commercial Area

6.10.5 商业设施
6.10.5 Commercial Facilities

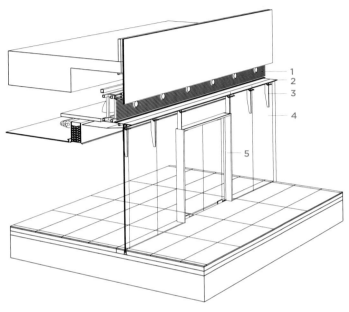

标准商业门头断面1
Typical Cross Section of Store Door Head 1

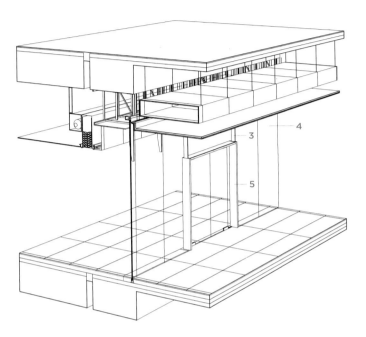

标准商业门头断面2
Typical Cross Section of Store Door Head 2

1. 银灰白色铝型材装饰百叶
2. 银灰白色铝板装饰门头压边
3. 夹胶钢化玻璃肋
4. 夹胶钢化超白玻
5. 不锈钢商业入口门框

1. Silver-Grey-White Aluminum Lamella
2. Silver-Grey-White Aluminum Plate Door Head Blank Pressing
3. Laminated Tempered Glass Fin
4. Ultra-Clear Laminated Tempered Glass
5. Stainless Steel Door Frame at Commercial Entrance

国际商业区
International Commercial Area

国内商业区
Domestic Commercial Area

6.10.6 登机桥固定端

结合混合机位布置、进出港主要流程、空侧设计条件、航站楼建筑设计条件等进行登机模式设计,设计与之相匹配的登机桥固定端。

1. 功能设计。针对国内混流、国际进出港、国际国内可转换三种进出港流程,合理规划功能与设施布局,实现资源利用与使用效率最大化。
2. 外形设计。延续了航站楼的设计手法,大胆对形体进行了斜切,结合立面铝板斜线分隔,延续了航站楼的动感造型特色,使之与航站楼主体有很好的体量过渡,也有效减少了空间体积而有利于节能。体现了现代、简洁、流畅、精致的建筑风格。外墙采用与航站楼模数相匹配的横明竖隐玻璃幕墙与铝板幕墙穿插,屋面采用檩条支承的铝镁锰金属屋面系统,在保障功能的前提下为旅客营造明亮、通透的视觉效果。
3. 空间设计。引入岭南建筑"敞厅"的建筑空间概念进行设计,把登机桥固定端当作一个独立建筑,结合其平面功能布局、流程设计、疏散要求对空间进行整合,在最高效解决功能需求的同时,创造其独特的通透、开敞、紧凑的整体空间效果。
4. 人性化设计。旋转平台标高实现与各类型飞机舱门标高与航站楼内各楼层标高的平顺对接,满足旅客的出行体验;可转换机位与国际机位登机桥固定端设有垂直电梯与自动扶梯,提升旅客出行体验,方便残疾人、行动不便的人进入飞机;登机桥固定端设有残疾人扶手等无障碍设施,地面坡度与防滑度均符合相关规范与标准。
5. 结构设计。整体采用钢桁架结构,为保证室内空间的通透,结构设计以高强度的吊杆代替柱子;在满足人行舒适度的前提下,为减薄楼梯厚度,楼梯采用中间高两边低的多梁并排形式;在登机桥的长度方向,采用高强度的拉杆代替桁架的斜腹杆,保证结构的稳定性及减少楼层的挠度;利用垂直电梯井道的结构一并作为主体结构的支撑杆件,减少过道的挠度,保证其舒适性;抗风柱与桁架上下弦间采用刚接连接,进一步降低上下弦即楼层梁的高度,减少楼层厚度。
6. 消防设计。建构完善的消防设施系统,其中自动喷淋灭火系统为国内首次采用。

6.10.6 Fixed Ends for Boarding Bridges

Based on the layout of swing stands, main arrival/departure flows, airside design conditions, and the architectural design conditions of the terminal, the design creates compatible fixed ends for the boarding bridge.

1. Functional design. Through reasonable planning of functional and facility layout in consideration of the three types of arrival/departure flows, i.e. domestic mixed flows, international arrival/departure flows, international and domestic switchable flows, the design realizes maximum resource utilization and service efficiency.
2. Appearance design. The appearance design continues the design approach of the terminal, with bold oblique cut for building shape and diagonal dividing lines for vertical aluminum plate on facade to continue the dynamic design of the terminal. This also offers smooth transition in volume from the main building, effectively reducing energy consumption due to less space and volume and showcasing modern, simple, smooth and delicate architectural style. The external wall is designed with glass curtain wall in horizontally exposed and vertically hidden frames interwoven with aluminum curtain wall that are compatible with the terminal modules. Al-MG-Mn metal roofing system supported by purlins is adopted to shape a bright and transparent environment for passengers while ensuring functionality.
3. Spatial design. The design introduces the spatial concept of Lingnan-style "open hall". By regarding the fixed ends of boarding bridges as an independent building, it integrates the spaces based on the planar functional layout, flow design and evacuation requirements, achieving unique effect of transparent, open and compact spaces on the whole while accommodating the functional demands in the most effective way.
4. People-oriented design. The elevation of the rotary platform enables smooth connection with the cabin doors of various types of aircrafts and various floors of the terminal, guaranteeing unobstructed travel experience for passengers. The fixed ends of boarding bridges for swing stands and international stands are equipped with elevator and escalator to better serve the disabled and those with physical inconvenience. At the fixed end, accessible facilities like accessible railings are provided and the surface slope and skid resistance all comply with relevant codes and standards.
5. Structural design. The design adopts steel truss structure. To ensure transparent interior space, high-strength booms instead of columns are adopted in structural design. To thicken the stairs without compromising walking comfort level, it adopts the layout of parallel beams higher in the middle and lower on both sides. Along the length of the boarding bridge, the design uses high-strength tie bar instead of diagonal web member for the truss to ensure structural stability and less floor deflection. It also uses the elevator shaft structure concurrently as a supporting rod for the main structure to reduce the passage deflection and ensure its comfort level; and adopts rigid connection between the wind-resistant column and the upper and lower strings of the truss to further reduce the height of upper and lower strings, namely the floor beam height, and the floor thickness.
6. Fire protection. The design establishes a complete fire protection facility system, including the automatic sprinkler system that is firstly used in China.

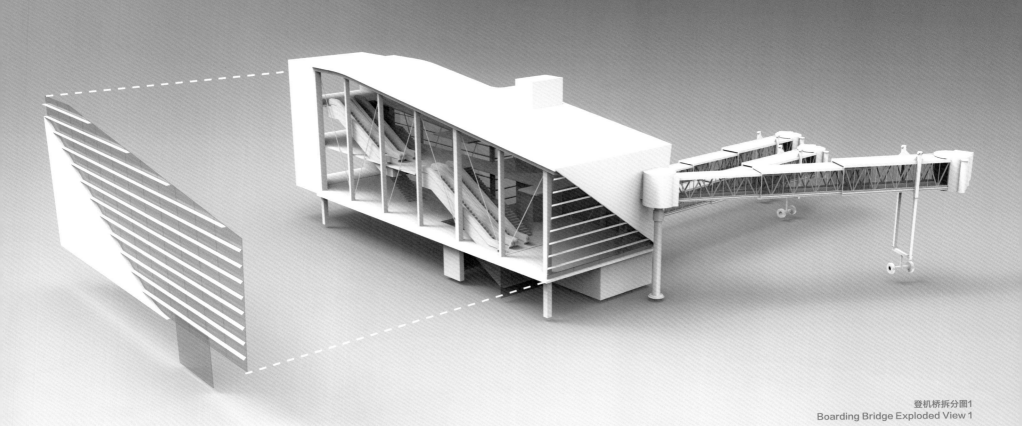

登机桥拆分图1
Boarding Bridge Exploded View 1

登机桥拆分图2
Boarding Bridge Exploded View 2

登机桥外景 1
Exterior View of Boarding Bridge 1

登机桥外景 2
Exterior View of Boarding Bridge 2

登机桥外景 3
Exterior View of Boarding Bridge 3

登机桥内景
Interior View of Boarding Bridge

登机桥外景 4
Exterior View of Boarding Bridge 4

柜台
Counter

座椅
Seat

6.10.7 柜台及座椅

柜台设计。柜台按照功能主要分为值机柜台、安检柜台、边检柜台、服务柜台、咨询柜台、登机柜台。外表面采用木纹及原色拉丝不锈钢面板，与二号航站楼室内风格协调统一；台面采用中国黑花岗石及酚醛树脂板，美观耐用；柜台尺寸结合人体工学精细设计，便于使用。

座椅设计。座椅按人体工程学原理设计，每个座位设计成独立的立体曲线结构；椅面、椅座整体为聚氨酯材料，内衬高强度铁骨架，绿色环保。椅脚、扶手采用铝压铸制作，美观、坚固且不生锈。脚底增设调整支撑，防滑耐磨。座椅根据不同功能空间选择了5种色系；每两个座椅中间安装一组USB充电插座，每组座椅配备一个单相五孔安全性插座。

6.10.7 Counters & Seats

Counters. As per functions, counters mainly include check-in counter, security check counter, immigration inspection counter, service counter, inquiry counter and boarding counter. The counters are cladded by brushed stainless steel panel in wood grain and primary color, harmonizing with the interior style of T2; their countertops use Chinese black granite and phenolic resin that are both aesthetic and durable; and their sizes are designed in human scale for easy use.

Seats. Designed as per ergonomic principles, each seat is in an independent solid curve structure. The seat surfaces and seats are made of high-strength iron skeleton wrapped by polyurethane materials, which are green and environmental-friendly. The legs and handrails made of aluminum die casting are aesthetic, solid and resistant to rust. The bottoms of the legs are provided with adjustable support for skid and abrasion resistance. The seats are made in five colors subject to different functional spaces. One USB charging socket is installed between every two seats and a single-phase five-hole safe socket is provided for each set of seats.

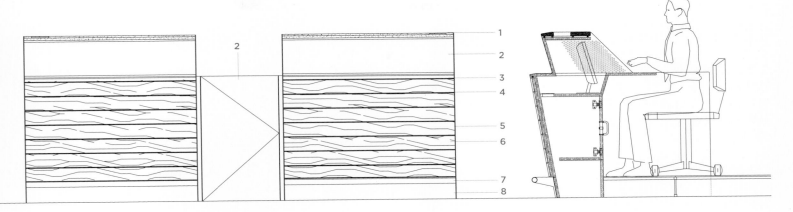

柜台正立面
Counter Front Elevation

柜台剖面
Counter Section

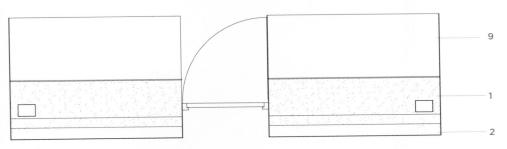

柜台平面
Counter Plan

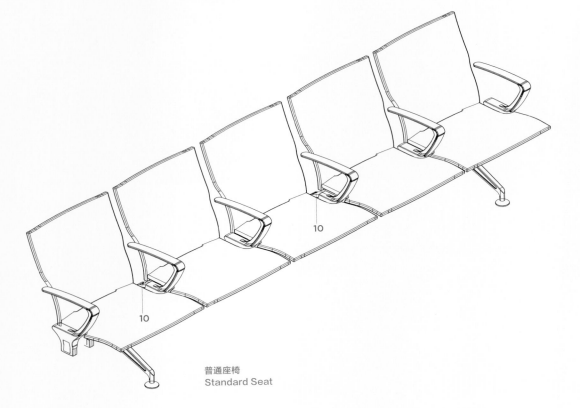

普通座椅
Standard Seat

1. 30mm厚中国黑花岗石
2. 2mm厚拉丝不锈钢面板,内衬18mm厚海洋胶合板
3. 35mm×5mm厚拉丝不锈钢防撞收边条
4. 15mm×5mm厚拉丝不锈钢防撞条
5. 4mm厚拉丝不锈钢侧面板
6. 18mm厚海洋胶合板,外贴1mm厚木纹酚醛树脂防火板饰面
7. 2mm厚Φ38拉丝不锈钢防撞杆
8. 1.5mm厚拉丝不锈钢踢脚板,内衬18mm厚海洋胶合板
9. 18mm厚海洋胶合板,面贴2mm厚酚醛树脂板
10. 座椅充电装置

1. 30mm Chinese Black Granite
2. 2mm Brushed Stainless Steel Plate Lined by 18mm Marine Grade Plywood
3. 35mmX5 mm Brushed Stainless Steel Anti-Collision Close-up Strip
4. 15mmX5 mm Brushed Stainless Steel Anti-Collision Strip
5. 4 mm Brushed Stainless Steel Side Board
6. 18 mm Marine Grade Plywood Covered by 1mm Wood-Grained Phenolic Resin Fireproof Plate
7. 2mmΦ38 Brushed Stainless Steel Bumper Bar
8. 1.5mm Brushed Stainless Steel Skirting Board Lined by 18 mm Marine Grade Plywood
9. 18 mm Marine Grade Plywood Covered by 2 mm Phenolic Resin Plate
10. Seat Charging Device

6.10.8 卫生间

合理规划卫生间布点，按照规范要求控制间距。航站楼内旅客在任一点到达最近卫生间距离不超过50m。卫生间男女计算比例为1:1，厕位数量按照区域聚集人数每50人设置一个大便器，男女大便器数量比例接近1:2；洗手盆数量按照每150人设置一个，女卫生间适当增加洗手盆数量。国内区域蹲便器坐便器比例为8:2，国际区域蹲便器坐便器比例为2:8。

卫生间的设计细节充分考虑人性化服务，采用简单的迷宫式设计，入口不设门，方便出入，同时实现视线遮挡。公共卫生间采用1.8m×1.2m厕格标准块，厕隔不使用时门保持15°向内开启状态，入口处布置1.2m×0.6m行李放置区，背后设置0.3m宽的手提行李置物台；小便器间距0.9m，并设置0.2m宽的手提行李置物台；洗手台标准间距0.8m，在使用便利性及布置效率方面取得了较好的平衡。洗手台流程设计充分考虑运营需求，每个洗手盆分别配置防滴落洗手液，每两个洗手盆之间配置一个擦手纸盒，方便就近擦手，避免洗手水大范围滴落。集中设置大型擦手纸回收箱，便于后勤管理。

卫生间设置风机盘管独立供冷，保证卫生间温度比外部公共空间低2~3℃。厕格配置了上下双排风系统，每个便器后方都就近设置了排风口，有效减少卫生间气味，并依据下排风系统加宽了厕格后方的检修通道系统。在卫生间厕格后方、小便器上方、洗手台上方天花均布置了反射式灯带，提高局部亮度及照明舒适度。卫生间全部采用冲洗阀系统，相比水箱系统更容易维修更换。

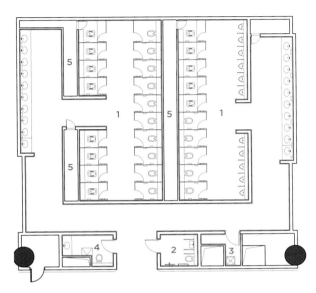

主楼标准卫生间平面图
Typical Plan of Washroom in Main Building

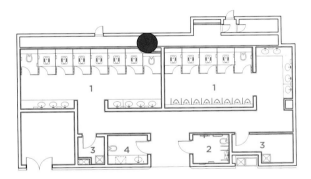

指廊标准卫生间平面图
Typical Plan of Washroom in Pier

1.普通卫生间
2.无障碍卫生间
3.清洁间
4.母婴室
5.检修通道

1. Standard Washroom
2. Accessible Washroom
3. Cleaning Room
4. Baby Care Room
5. Maintenance Access

6.10.8 Washrooms

Washrooms are planned in proper intervals as per codes and requirements. The distance between any location in the terminal and the nearest washroom is no more than 50 m. The ratio of male to female washrooms is 1:1. One closet pan is provided for every 50 people in view of population size and the ratio of male to female closet pans is close to 1:2; one basin is provided for every 150 people, with a bit more in female washrooms. The ratio of squatting pans to pedestal pans is 8:2 in domestic service area and 2:8 in international service area.

With full consideration of people-oriented service, washrooms are in simple labyrinth design with no doors at the entrance, which guarantees easy access while ensuring sightline obstruction. Typical cubicle blocks of 1.8 m X 1.2 m are adopted in public washrooms, where cubicle doors stay open inwards by 15° when unoccupied. Baggage area of 1.2 m X 0.6 m is provided at the entrance with a hand baggage counter of 0.3 m wide at the back. Urinals are placed in an interval of 0.9 m and equipped with a hand baggage counter of 0.2m wide. Washbasins adopt a standard interval of 0.8 m, convenient in use and efficient in layout. The flow design of washbasins takes full consideration of operational needs, namely, each washbasin is equipped with an anti-dropping hand cleaner, and every two washbasins are provided with a tissue case between them for convenient hand wiping without wide-range dropping. Big recycle bins for used paper towels are centralized for the ease of BOH management.

Washrooms are designed with independent fan-coil cooling system to ensure temperature of 2-3 °C lower than exterior public space. Their cubicles are provided with upper and lower exhaust systems, with exhaust vent behind each closet pan for effective odor release and the maintenance access system widened at the back of the cubicles in consideration of the lower exhaust system. Reflective light strips are provided on the ceilings above cubicle backs, urinals and washbasins for brighter and more pleasant illumination. All washrooms are equipped with flushing valve system that is easier to maintain and replace than water tank system.

6.10.9 贵宾室

以提供功能齐全、舒适的一站式贵宾服务为设计理念。贵宾室设置专用出入口和停车场，并紧邻空侧服务车道边，实现贵宾专用流程的独立性与便捷性，同时设置连通普通旅客流程的竖向交通，为贵宾提供多种流线可能。结合岭南地域文化特点，打造极具特色的文化体验空间和尊贵的仪仗式接待空间。

6.10.9 VIP Room

The design aims to provide pleasant one-stop VIP services with complete functions. The VIP room is designed with dedicated access and parking lot in proximity to the airside service curbside to ensure independent and convenient VIP flow. It is also equipped with vertical circulation system leading to ordinary passenger flow area to allow for more options for VIPs. Featuring the cultural characteristics of Lingnan region, the VIP room offers distinctive cultural experience and distinguished clientele reception.

国内贵宾室平面图
Domestic VIP Room Plan

1. 国内政要贵宾门厅
2. 国内商务贵宾门厅
3. 展示区
4. 政要贵宾出发安检通道
5. 公共休息区
6. 商务贵宾到达通道
7. 商务贵宾出发安检通道
8. 行李托运办理柜台
9. 行李库
10. 政要贵宾室
11. 商务贵宾室

1. Domestic Political VIP Foyer
2. Domestic Business VIP Foyer
3. Display Zone
4. Security Check Channel for Departure Political VIPs
5. Public Sitting Area
6. Business VIP Arrival Channel
7. Security Check Channel for Departure Business VIPs
8. Baggage Check-In Counter
9. Baggage Depot
10. Political VIP Hall
11. Business VIP Room

贵宾室
VIP Room

贵宾室门厅
Foyer of VIP Room

6.10.10 其他服务实施：母婴候机室、儿童娱乐区、无障碍设施等

1. 母婴候机室

母婴室的选址及设计参考了《广州市公共场所母婴室建设指导手册》，并与业主及使用单位经过多次的论证，其设计特点如下：

（1）选址在人流动线的主节点上。东、西、北三个指廊办票大厅及混流大厅均有设置母婴候机室，满足带婴儿的女士旅客需求。

（2）配备多种功能设施，如暖奶器、直饮水机、电视机、烘干机；布置人性化的功能区，如亲子洗手间、哺乳区、换尿布区、玩具兼阅读区、婴儿爬行区等。

（3）配备良好的通风换气设备，确保为室内提供质量良好的空气。

（4）门口有足够明显的引导标识，引导用户到达母婴室。

2. 儿童娱乐区

每个指廊选择合适位置设置儿童娱乐区，满足家长更方便照顾儿童旅客的需求，提升机场人性化服务。

3. 无障碍设施

（1）二号航站楼前设有红绿灯的路口和盲人过街音响设施。

（2）无障碍安检通道。采用无障碍安检验证台。

（3）盲道。人行道盲道从出发车道边引导至楼内问询柜台位置；楼内电梯、自动步道、扶梯及楼梯设置有盲道提醒设施。

（4）无障碍卫生间。每处卫生间配备一个独立的无障碍卫生间，设置了电动推拉门、无障碍坐便器、无障碍洗手盆和满足使用轮椅旅客使用的墙面倾斜镜子。

（5）无障碍登机桥。登机桥内设置满足残疾人使用的坡道、扶手、电梯。

（6）无障碍车位。无障碍车位占总车位比例为2.1%。

（7）无障碍电梯。二号航站楼客梯轿厢内设置扶手和带盲文的选层按钮；轿厢上、下运行及到达有清晰显示和报层音响。

（8）无障碍座椅。按照座椅数量的2%设计，在座椅靠近主要交通走道的位置设置。

6.10.10 Others: Baby Care Waiting Rooms, Children Entertainment Areas, Accessible Facilities

1. Baby Care Waiting Rooms

The location and design of baby care rooms references the *Guidelines for the Development of Baby Care Rooms in Public Spaces of Guangzhou* and has been repeatedly justified by the Client and occupants. They have the following design characteristics:

(1) Placed at the main nodes of passenger flows. They are provided in the check-in halls of the east, west and north piers as well as the mixed-flow hall to serve female passengers with babies.

(2) Equipped with multiple functions. They are equipped with milk warmer, direct water dispenser, television and dryer, and comprise several people-oriented functional areas like

母婴候机室
Mother and Child Waiting Room

儿童娱乐区
Children Entertainment Area

parent-child washroom, lactation area, diaper-changing area, toys & reading area, baby crawling area, etc..

(3) Good ventilation. They are equipped with good ventilation facilities to ensure quality air indoors.

(4) Eye-catching signage. They all have prominent signage at the entrance for easy orientation.

2. Children Entertainment Areas

Children entertainment area is placed at an appropriate position of each pier for parents to take better care of their children, adding to high service level of the airport.

3. Accessible Facilities

(1) An intersection with traffic lights and acoustic crossing device for visually impaired people are provided in front of T2.

(2) Accessible security check channel. Accessible verification counter for security check is adopted.

(3) Blind tracks. The blind track on the sidewalk can lead visually impaired passengers from the curbside to the inquiry counter inside the building; signs for blind tracks are provided at elevators, automatic people movers, escalators and stairs.

(4) Accessible washrooms. Each washroom has one independent accessible toilet equipped with electric sliding door, accessible pedestal pan, accessible washbasin and tilting mirror on the wall for wheelchair users.

(5) Accessible boarding bridges. Accessible ramps, railings and elevators are provided inside the boarding bridge.

(6) Accessible parking. Accessible parking spaces account for 2.1% of the total parking spaces.

(7) Accessible elevators. Railings and buttons with braille for floor selection are provided inside the passenger elevators of T2; clear display and floor-reporting audio device are provided to remind passengers of upward/downward running or arrival.

(8) Accessible seats. 2% of the seats are accessible, placed near main walkways.

过街红绿灯
Traffic Lights

无障碍车位
Accessible Parking Space

6.11 景观设计 — 6.11 LANDSCAPE ARCHITECTURE

结合绿色航站楼的整体空间布局组织整体景观体系

1. 在二号航站楼安检厅、联检厅北侧,与北指廊之间营造岭南花园,让旅客在现代化的航站楼内可以感受到传统园林的魅力,是人与自然融合的最佳实践,身处这里,犹如漫长旅途的一片绿洲,体现了广州地处中国南方的气候特点。岭南花园属于狭长式用地,设计采用斜线条式的铺装布局,使空间变得生动有趣,配以造型树池、水景、汀步等穿插布局,以达到小中见大,空间层次分明的效果。北指廊的国际到达旅客可以欣赏到岭南花园的美景,对广州留下美好的第一印象。

2. 在交通中心及停车楼建筑外立面上采用垂直绿墙,屋顶设置绿化停车场,在行车道和停车位旁以秋枫为主要树种,局部点缀开花植物大腹木棉,与二号航站楼连通处水景种植棕榈树种银海枣,体现绿色南国风情。

A holistic landscape system based on overall spatial layout of green terminal

1. Lingnan gardens are provided between the north pier and the north side of the security check hall and CIQ hall of T2. It allows passengers to experience the charm of traditional gardens in a modern terminal setting. This is the best practice to integrate man and nature. Being here is like encountering an oasis during a long journey, and passengers can experience the climatic features of Guangzhou as a city in southern China. The land for Lingnan gardens is narrow and long, with the design of oblique lines on pavement to create vivid and intriguing spaces. Coupled with tree pool, waterscape, stepping stones on water surface and other design, much can be reflected in small details and the orderly spaces. International arrival passengers from the north pier can enjoy the beautiful scenery of Lingnan gardens and make a good first impression on Guangzhou.

2. The design provides vertical green walls for the façades of GTC and parking building and greening parking lots on their roofs. Along the vehicular lanes and by the parking spaces, bischofia javanicas interspersed with flowering plant ceiba speciosas are planted. By the waterscape at the connection with T2, phoenix sylvestris of palm tree species typical of Southern China charm are planted.

二号航站楼绿化平面
T2 Greening Plan

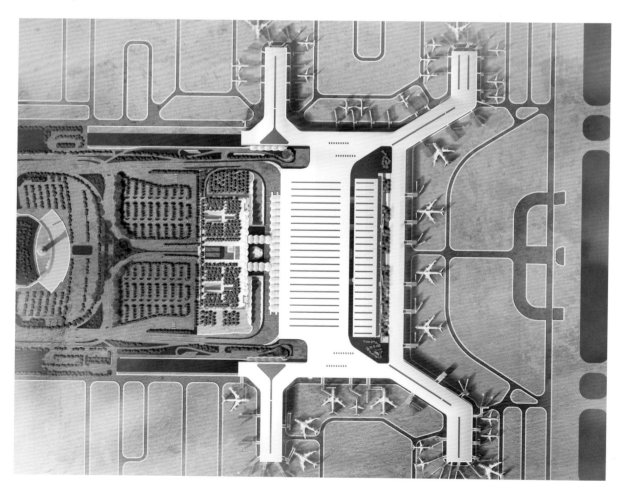

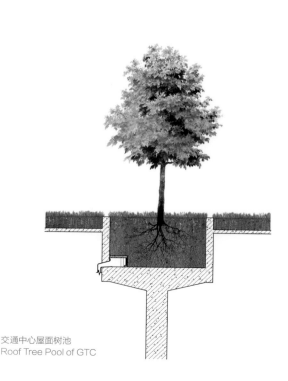

交通中心屋面树池
Roof Tree Pool of GTC

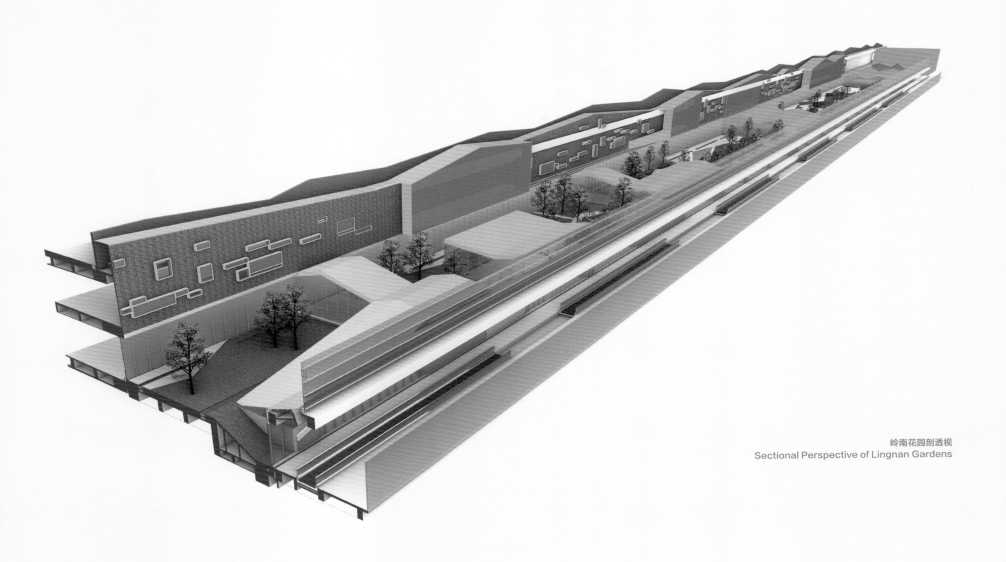

岭南花园剖透视
Sectional Perspective of Lingnan Gardens

岭南花园1
Lingnan Gardens 1

岭南花园2
Lingnan Gardens 2

6.12 绿色建筑设计

6.12.1 设计目标与分析

6.12.1.1 气候特点
夏热冬暖地区全年日照时数长，辐射强烈，应着重考虑建筑遮阳体系，以降低建筑室内太阳辐射的热量，进而降低建筑空调能耗。全年平均风速在2m/s左右，夏季及过渡季主导风向为东南风，冬季为北风，自然通风潜力较大，建筑应考虑最大可能的再过渡季利用自然通风。通过对气候特点的分析可以得到在这一区域适用的绿色建筑策略，根据这些策略寻求建筑设计上的实现方式和采用的技术措施。

6.12.1.2 航站楼建筑特点
航站楼是机场的标志性建筑，占地、建筑规模及体量较大；功能和流线复杂；实墙面较少，玻璃幕墙较多，窗墙比较大；室内多为高大空间；建筑运行时间长；内部人流量大；对室内环境品质要求较高。全年耗能量大，运行费用高。航站楼能耗主要就为中央空调及照明，另外还有很多传送设备、弱电设备等。通过对建筑特点的分析在绿色建筑设计中应重点考虑航站楼的节能策略与技术措施。

6.12.1.3 务实的绿色建筑设计目标
广州白云国际机场作为国内三大枢纽机场之一，体现着整合区域交通、集约城市空间、打造城市节点、贯彻绿色生态，落实环境品质等城市可持续发展理念。根据我国绿色机场发展的阶段，项目应打造成为我国华南地区典型绿色机场建筑，因此项目绿色建筑设计目标确定为国家三星级绿色建筑，根据《绿色建筑评价标准》GB50378-2006开展绿色建筑设计。

6.12.2 设计策略

6.12.2.1 舒适微气候环境的营造
1. 绿地系统规划。充分利用航站楼陆侧室外场地，在航站楼陆侧建设围绕航站楼、交通中心及进出港高架路的大面积绿地。场地硬质铺装尽可能采取透水性铺装，项目室外透水地面面积比达到44%。
2. 立体绿化系统。在安检区、联检区北侧设置岭南花园；交通中心结合空间布局、结构体系设置屋顶绿化，立面外墙设置层次丰富的墙面绿化，形成完整的立体绿化系统，屋面绿化比例达到32.5%。
3. 室外遮阳设计。利用航站楼大屋面挑出部分、室外PTFE膜结构体系对航站楼陆侧硬质地面和不可透水路面提供遮阳，减少地面对太阳辐射的吸收，改善微气候环境，降低周边环境空气温度。
4. 自然通风设计。通过在外窗和玻璃幕墙设置可开启部分以充分利用自然通风，同时屋面设置气动排烟窗，以促进建筑过渡季利用自然通风。同时对位于航站楼内部联检区等通风条件较差的区域，通过设置可供旅客休憩的岭南花园以加强区域的自然通风条件，改善建筑用能效率和提高室内空气品质。

绿地系统规划
Green Space System Planning

车道边张拉膜雨甘
Curbside Tensile Membrane Canopy

6.12 GREEN BUILDING

6.12.1 Design Objective and Analysis

6.12.1.1 Climatic Characteristics
With hot summer and warm winter, the project region features long sunshine duration and strong radiation all year round, so the design should give consideration to building sun-shading system to lower interior heat from solar radiation and decrease AC energy consumption. In Guangzhou, the average annual wind speed is 2m/s or so with prevailing southeast wind in summer and transition seasons and north wind in winter. This brings great potential for natural ventilation. The design should make greatest efforts to realize natural ventilation in transition seasons. Through the analysis of climatic characteristics, green building strategies applicable in this area can be concluded, based on which feasible and adoptable technical measures can be worked out.

6.12.1.2 Architectural Characteristics
T2 is the landmark of Baiyun Airport. It features large footprint, building size and volume; complicated functions and flows; more glass curtain walls than physical walls hence big window-wall area ratio; mostly tall interior spaces; long building operation time; and big pedestrian flows. It requires high interior environmental quality, hence large energy consumption throughout the year and high operation costs, which mainly comes from central AC and lighting and other transmitting and low-voltage equipment. Based on analysis of these architectural characteristics, the green building design of T2 focuses mainly on energy efficiency strategies and technical measures.

6.12.1.3 Pragmatic Objective
As one of the three major hub airports of China, Baiyun Airport embodies urban sustainable development concepts, including regional transportation integration, urban spatial intensity, urban symbol creation, and green ecological and environment quality implement. Based on the green airport development status in China, the Project should be developed into a typical green airport building in South China, hence the objective of pursuing national 3-star green building design label. The green building design of the Project is carried out in accordance with Assessment Standard for Green Building GB50378-2006.

6.12.2 Design Strategies

6.12.2.1 Comfortable Microclimate
1. Green space system planning. The landside outdoor ground of T2 is fully utilized to provide extensive green spaces around T2, GTC and the arrival/departure viaduct. Permeable rigid pavement is adopted to the greatest extent. The proportion of outdoor permeable surface of the Project reaches as high as 44%.
2. Vertical greening system. Lingnan gardens are provided on the north side of security check and CIQ areas; roof greening and hierarchical wall greening are provided for the GTC in consideration of its spatial layout and structural system, forming a complete vertical greening system. The proportion of roof greening reaches 32.5% in the Project.
3. Outdoor sun-shading design. Sunshade area is provided on the Terminal landside made from rigid materials and impermeable road surface by selecting some outdoor PTFE membrane structural system on the large roof of Terminal, so as to reduce the road surface's absorption of solar radiation, improve microclimatic environment and lower air temperature in surrounding environment.
4. Natural ventilation design. The design provides operable sash in external windows and glass curtain walls to maximize natural ventilation, and pneumatic smoke exhaust window in the roof to facilitate natural ventilation in transit seasons. At the same time, for areas with poor ventilation such as the internal CIQ area in the terminal, atrium gardens where passengers can take a rest are provided to improve natural ventilation, building energy efficiency and indoor air quality.

立体绿化系统
Multi-level Greening System

6.12.2.2 采光与遮阳的平衡

1. 自然采光设计与优化。航站楼采用天窗采光和侧向玻璃幕墙相结合的方法改善和调节室内自然采光效果,办票大厅和安检大厅上空屋面分别设置22个3m×126m和3m×45m的带形采光天窗,每年节约用电量约280万度,约占照明总用电量20%。交通中心及停车楼设置2个26m×42m的采光天井,采光模拟结果显示建筑地下一层空间采光系数不低于0.5%的建筑面积占地下一层总建筑面积的比例达到5.3%。

2. 遮阳设计与优化。在二号航站楼西五、西六指廊及连接指廊的西向玻璃幕墙设置机翼型可调节电动遮阳百叶,有0°、30°、60°和90°四个角度。外遮阳系统的控制纳入航站楼建筑的建筑设备监控系统(BAS),根据工作时间表或昼夜、季节,自动控制百叶至对应的预设角度。在建筑东向采用室内电动百叶卷帘遮阳系统。

6.12.2.3 可再生能源的利用

1. 太阳能光电系统设计。航站楼金属屋面设置用户侧并网分布式光伏发电系统,光伏系统安装于标高为31.2m的安检大厅顶部屋面,装机总容量2兆瓦。相当于每年节约标煤779.8t,减排CO_2量2043.12t,SO_2减排量6.63t,氮氧化物5.77t。选用的光伏组件表面镀有吸光材料,采用专用的超白玻璃,其核心部分有陷光结构,透光率可达91.5%,反射率低于3%,能够有效吸收太阳光,减少光污染。

2. 太阳能光热系统设计。采用太阳能集中热水系统,为计时旅馆、头等舱及商务舱区域提供生活热水,辅助热源由空气源热泵提供。具体技术措施为在主楼东、西侧屋面各设置一套太阳能加热系统,系统采用平板型太阳能集热器,集热器总面积846m^2,日太阳能热水(60℃)用量49.4m^3,占航站楼总生活热水用量75.3%。

6.12.2.4 结构体系的优化

1. 超长大跨度预应力混凝土结构。采用18m大跨度框架结构及分缝技术,巧妙地设计为预应力混凝土双梁结构,避免预应力穿框架柱,加快了工期,节约了成本。

2. 单跨预应力混凝土柱排架结构。对支撑钢结构屋盖的混凝土柱施加预应力,柱截面尺寸得到优化,建筑效果和经济效益显著。

3. 空心板结构。航站楼首层楼板局部采用大跨度空心板结构,内置预应力筋,合理控制了裂缝,大大节约了建筑空间和成本,加快了施工工期。

4. 抽空网架结构和钢管混凝土柱。支撑屋盖柱采用钢管混凝土柱,钢管为高强度Q345B级钢材,管内混凝土为C50级高强混凝土,与上部钢结构更能协调变形。网架采用加强肋布置及抽空处理,网架用钢量约50kg/m^2,做到了轻省,结构体系合理。

经过上述结构体系优化,项目可再循环材料使用比例达到11.93%。

6.12.2.5 雨水回收利用

项目采用雨水收集利用技术,建地下储存调节水箱,雨季收集利用雨水,旱季及用水量不足时以机场污水处理厂产生的中水补充,提供区域内绿化浇灌、道路冲洗、幕墙清洗用水。项目雨水设计主要收集航站楼屋面雨水,总收集面积为约46000m^2,项目非传统水源利用率达到2.46%。

6.12.2.6 高效设备系统

1. 空调采暖系统。项目采用集中式水冷中央空调系统,仅考虑夏季制冷,不考虑冬季采暖。项目冷水机组能效比COP、水泵输送能效比、风机单位风量耗功均满足《<公共建筑节能设计标准>广东省实施细则》DBJ15-51-2007的要求。

2. 照明系统。采用放射与树干相结合的配电方式,分区设置配电间。各功能区域结合空间尺度与功能布局特点进行灯具及电器附件选择。采用智能照明控制系统对航站楼和交通中心的照明进行集中控制和管理。各房间或场所的照明功率密度值不超过《建筑照明设计标准》GB50034规定的现行值。

3. 节水系统。所有用水器具均采用符合国家标准的节水型卫生器具,进行用水分项计量,在与航站区室外给水管连接的引入管处设给水总表,主楼、每条指廊设分总表,分总表后根据管理需求设分表。室外绿化灌溉采用喷灌等节水灌溉系统。

4. 项目还采用能耗分项计量系统、建筑自动监控系统、室内空气质量监控系统等综合节能系统。

屋面天窗
Roof Skylight

可调节电动遮阳百叶
Adjustable Electric Sunshading Louvers

6.12.2.2 Balance between Daylighting and Sun-shading
1. Daylighting design and optimization. Skylights combined with lateral curtain walls are adopted for T2 to improve and regulate interior daylighting effect. That is, 22 skylight strips sized 3mx126 m and 3mx45m are respectively provided in the roof of the check-in hall and security check hall. They can save electricity consumption of about 2.8 million kilowatts a year, which accounts for 20% of the total power consumption for lighting. Two daylighting patios sized 26mX42m are provided in GTC and the parking building. Lighting simulation shows that the floor area of B1 with lighting coefficient of no less than 0.5% accounts for 5.3% of the total floor area of B1.
2. Sun-shading design and optimization. Wing-type adjustable electric sun-shading louvers are provided in the western glass curtain walls of the West 5th Pier, West 6th Pier and connecting pier, allowing for operable angels of 0°, 30°, 60°and 90°. Control of exterior sun-shading system is incorporated into the Building Automation System (BAS) of T2, which can automatically control the louvers to corresponding preset angle as per working timetable or day/night/season. In the eastern façade, interior electrical louver sun-shading system is adopted.

6.12.2.3 Utilization of Renewable Energy Sources
1. Solar electricity system. Terminal metal roof provides user-side grid connection and distributed power generating network. Photovoltaic system is placed on the roof of the security check hall at Level 31.2m with a total installed capacity of 2 MW, which equals to annual standard coals conservation of 779.8 tons, CO_2 emission reduction of 2,043.13 tons, SO_2 emission reduction of 6.63 tons and NOx of 5.77 tons. The selected photovoltaic module uses light absorbing material on the surface and dedicated ultra-white glass with light trapping structure in the core that can realize light transmittance of 91.5% and reflectivity rate of lower than 3%, effectively absorbing sunlight and reducing light pollution.
2. Solar thermal system. Domestic hot water for hourly charged hotel, first-class and business-class cabins are provided by adopting solar centralized hot water supply system with supplementary air source of heat pump. To be specific, place a set of solar heating system on the eastern and western roof of the main building respectively. The system includes a flat-plate collector, totaling an area of 846 m^2. The solar hot water (60 ℃) consumption per day is 49.4m^3, accounting for 75.3% of the total domestic hot water consumption in the Terminal.

6.12.2.4 Optimization of Structural System
1. Super long span pre-stressed concrete structure. The design cleverly adopts 18 m long span frame structure and parting technology to provide pre-stressed concrete double-beam structure that can prevent prestress from passing through the frame columns, shortening the construction duration and saving the costs.
2. Single span pre-stressed concrete column bent structure. Prestress is applied to the concrete columns supporting the steel roof. This optimizes the sectional dimension of the columns, contributing to prominent architectural effect and economic efficiency.
3. Hollow slab structure. Long span structure with hollow slab is adopted on the first floor of T2. Inside the slab structure pre-stressed steel is provided, which controls the crack in a rational manner, greatly saves the building space and costs and accelerates the construction.
4. Vacuum grid structure and concrete filled steel tubular columns. Concrete filled steel tubular columns using Q345B steel are used to support the roof. The tubes are filled with C50 high-strength concrete to better coordinate with the upper steel structure. The grid adopts reinforcing ribs and vacuum treatment with steel consumption of about 50 kg/m^2, making the structural system lighter and more cost effective. With the aforementioned optimization, the proportion of recycled materials in the Project reaches 11.93%.

6.12.2.5 Rainwater Recycling
Rainwater harvesting and utilizing technology is adopted in the Project. Underground water storage and regulating tanks are provided to harvest rainwater in rainy seasons for greenery irrigation, road flushing and curtain wall cleaning. In dry or water shortage seasons, reclaimed water produced by the sewage disposal plant in the airport will serve as supplement for the aforesaid uses. The rainwater recycling system of the Project harvests mainly rainwater on the terminal roof covering a total surface area of about 46,000m^2. The Project's utilization rate of non-traditional water sources is 2.46%.

6.12.2.6 Efficient MEP System
1. HVAC system. Centralized water-cooled central AC system is deployed for cooling in summer without heating in winter. The COP, ER and WS of the water chilling unit of the Project all meet the requirements of Detailed Rules of *Guangdong on Implementing the Design Standard for Energy Efficiency of Public Buildings DBJ15-51-2007*.
2. Lighting system. Radiation-trunk integration is adopted to distribute power with distribution rooms provided by zone. Lamps and electrical accessories in functional zones are selected in consideration of spatial scale and functional layout. Smart lighting control system is adopted for centralized management of lighting in T2 and GTC. Lighting power density values of all rooms or places are controlled below the existing ceiling stipulated by the *Standard for Lighting Design of Buildings GB50034*.
3. Water efficiency system. All watering devices comply with national standard for water efficiency sanitary fixtures. Water consumption is measured by items. A general water meter is provided at the inlet pipe connected with outdoor water supply pipe of the terminal, sub-general water meters are provided at the main building and every pier, and sub meters are allocated based on management demands. Spray irrigation and other water-saving irrigation system are adopted for outdoor greening.
4. Other comprehensive energy efficiency systems of the Project include energy consumption sub-metering system, building automatic monitored control system, and interior air quality monitoring and control system.

第7章

技术设计
CHAPTER 7 TECHNICAL SYSTEMS

7.1　行李系统设计
7.2　结构设计
7.3　机电设计
7.4　市政工程设计

7.1　BAGGAGE
7.2　STRUCTURE
7.3　MEP
7.4　UTILITIES

7.1 行李系统设计

7.1.1 概述

本工程行李处理系统以实现最高的运输效率、安全性、可靠性为目标。为了适应未来可能的需求量增长,具备足够的灵活性和扩容能力。

以满足2020年4500万人次年旅客吞吐量为设计目标,考虑高中转比例,从旅客流程着手,迎合航空公司枢纽运作与旅客出行便利需求。

7.1.2 设计特点

采用国内首个凭借托盘DCV技术实现分拣的机场行李系统,该系统较传统输送机翻盘式分拣机系统在功能和技术上都有极大的提高。主要具有以下特点:

1. 采用DCV小车自动分拣系统,运行速度快、分拣时间短、效率高、准确率高,同时DCV自动寻址功能提升了行李系统运行的可靠性。
2. 设置高效便捷的中转行李处理系统,匹配助推二号航站楼枢纽功能的发展。
3. 应用RFID技术提升行李跟踪率,运行后追踪率保持在99.91%左右,确保安检和海关拒绝的行李的下线检查。
4. 建设自助行李托运系统,提升旅客体验及运行效率。集中设置全自动行李自助交运柜台,自助托运设备将按匹配双通道安检机的传统柜台尺寸进行设计,在未来逐步取代现有的传统值机柜台。
5. 建设大容量早到行李存储系统(EBS),为早到或需长时间停留的旅客先行将行李通过值机或中转送到行李处理储存系统存储。
6. 模块化设计,不但减少了行李系统的备品备件种类,还保证了DCV系统有更好的扩展性,日后系统更换与扩容改造会更加方便,降低不停航施工的风险。
7. 为一号航站楼行李系统接入预留接口,满足白云机场一、二号航站楼之间未来可能的中转行李处理或一体化运作需求。

7.1.3 行李处理流程

7.1.3.1 行李系统处理原理

行李系统包含始发行李处理系统、到达行李处理系统、中转行李处理系统、早到行李处理系统、大件行李处理系统、贵宾行李处理系统等。

7.1.3.2 始发行李处理系统

始发行李处理系统行李在值机柜台收集,经过行李安检系统、值机柜台行李输送线、DCV系统、早到行李存储系统(仅早到行李进入)、人工编码站,以及出发装运转盘,最终通过行李车收取并运送至指定航班的飞机。

7.1.3.3 到达行李处理系统

到达行李中的标准行李,由行李拖车运送到首层行李机房到达行李卸载区,行李被卸载到到达行李装卸线上,随后通过行李地沟输送线将行李导入到到达行李提取转盘。到达行李提取转盘有长(约100m)和短(约70m)两种类型,长转盘对应有2条输送线,短转盘对应一条。行李提取大厅共设有21个标准斜面行李提取转盘供旅客提取行李,其中国内11个,国际10个。

到港超大行李通过到达超大行李装卸线输送到位于旅客到达区域的超大行李到达柜台。特殊行李以及超大行李输送线无法运输的行李(例如帆板),由人工送往超大行李到达柜台。

7.1.3.4 中转行李处理系统

根据中转流线,中转行李处理系统分后台中转行李和再值机中转行李两大类。

1. 后台中转行李

航空公司通程联运中转业务旅客不需要提取行李,行李在后台行李机房进行中转处理。二号航站楼行李系统设置了6条中转行李输入线路,分布在行李分拣房的中部区域,类型包含:国内转国际、国际转国内、国际转国际、国内转国内。

各中转行李由行李拖车运送至行李机房中转线装卸处,行李卸载后输入中转行李输入线。除国内转国内的输入线路外,各中转行李输入线皆衔接1台安检机,以便可以安全检查所有与国际相关的中转行李。被拒绝的可疑行李,将由DCV系统分拣至相应查验区进行旅客开包检查。通过安检合格的行李,将分拣至所属的转盘。由于设计中已将国际国内区域的主分拣系统连接贯通,因此理论上国际转国内或国际转国际或国内转国际的中转行李皆可在任何一条国际国内中转系统的输入线装中转行李。

2. 再值机中转行李

对于无法获得通程联运政策支持的中转航班,旅客需在到达区提取行李并办理值机业务。为更好地服务此类旅客,行李系统亦配合流程在到达出口设置了2个再值机区,类型包含:国际转国内、国内转国际、国内转国内。其行李处理与始发行李相同。联程拒绝的行李,将由DCV系统分拣至相应查验区进行检查。通过检查合格的行李,将分拣至所属的转盘。

7.1.3.5 大件行李处理系统

1. 出港大件行李

出港大件行李走大件行李专用处理流程,由人工处理。旅客在国内、国际大件行李值机柜台办理行李托运,安检拒绝的行李现场开包检查处理,安检合格则由人工通过垂直电梯送往行李房分拣机房处理。

2. 进港大件行李

进港大件行李由行李拖车运至大件行李到达卸载线卸载,行李通过到达大件行李输送线输送到位于旅客到达区域的超大行李提厅。特殊行李以及超大行李输送线无法运输的行李(例如帆板),由人工送往超大行李到达柜台。

3. 中转大件行李

中转大件行李走大件行李专用处理流程,由人工处理。中转大件行李由行李拖车运至行李分拣机房大件行李中转安检机进行安检,安检合格则进入大件行李出发流程。

7.1.3.6 早到行李处理系统

二号航站楼国内、国际区分别设置了两组各2000个存储位的货架式早到行李存储系统EBS,远期可扩展至不少于6000件存储位。特殊情况下国内、国际区域可相互共享存储资源。早到行李的按随机分配的原则储存到货架内的存储位。EBS除了提供早到行李存储功能以外,同时也兼备作为第二级的空托盘缓存。

早到行李可源自始发或中转行李,源自始发行李将由预分拣子系统将早到的行李输送至早到储存处,源自中转行李,将送入主分拣系统,由分拣系统分拣至早到储存系统储存。早到行李采用全自动DCV货架存储系统。每架堆垛机的处理量最低为每一分钟100件。

7.1.3.7 贵宾行李处理系统

贵宾行李在整个行李系统负荷中只占非常小的比例,并且通常每次的处理量也比较少。设计为独立子系统,并不衔接到总行李系统。

出港贵宾行李在贵宾室的值机柜台处托运,行李通过单独的X光机进行安检,合格后收集起来用行李拖车送至对应的航班。

进港贵宾行李由行李拖车直接拖至贵宾行李到达区,然后人工送至贵宾休息室。

7.1.3.8 行李系统预留

所有远期系统扩建接口位置的本期设备,都预留与接口机械尺寸相同的整台设备或整数台设备。原则上,DCV系统任意位置的整台或整数台输送机,机械上可直接替换成水平45°分流器与/或90°侧向分流器,无需对接口上下游的设备做任何机械修改。未来可能的行李系统的扩建内容有:

1. 增设DCV分拣环路;
2. 增设集中CT安检区;
3. 扩建空托盘存储系统ECS,将ECS的存储能力从

2000件扩充至3000件;
4. 扩建早到行李存储EBS,将EBS的存储能力从4000件扩充至6000件。

7.1.4 行李安检系统

7.1.4.1 始发出港及中转再交运行李

为了符合民航局标准以及更有效地处理与旅客在场开包的安检模式,二号航站楼始发出港及中转值机再交运行李采用分散柜台式安检模式。柜台式安检模式对应的五级安检为:柜台式单/双通道安检机为一级安检设备,生成图像后标识有嫌疑的区域;二级为安检操作员100%判读一级图像,二级拒绝的行李传送到值机岛末端的开包间;进入开包间的行李一般直接进入第五级开包检查程序;有高危违禁品的行李则会先经过三级CT检查以及四级ETD设备人工处理。

行李安检的流程是将交运行李先经过值机双通道安检机,由安检人员远程对双通道安检机图像进行判读,安检不合格的行李送往开包房进行爆炸物跟踪探测(ETD)然后与旅客进行第五级行李开包检查。

7.1.4.2 后台中转行李

二号航站楼后台中转的托运行李,采用集中式五级安检模式。五级安检为:双视角高速安检机为一级安检设备,生成图像后标识有嫌疑的区域;二级为安检操作员100%判读一级图像,二级拒绝的行李由BHS传送到指定的检查站;进入检查站的行李一般直接进入第五级开包检查程序;有高危违禁品的行李则会先经过三级安检设备复检以及四级ETD设备人工处理。

后台各中转行李输入线(国内转国内除外)前端皆衔接设置1台安检机检查所有与国际相关的中转行李,行李先经过安检机,由安检人员远程对安检机图像进行判读,安检不合格的行李将由DCV系统分拣至相应检查站进行第五级旅客开包检查。

7.1.5 行李系统应急设计

7.1.5.1 值机收集主皮带应急

值机收集皮带之间设有跨越线,在任何一条值机收集皮带或DCV线发生故障时,行李将从原来的皮带通过水平分流器跨越到另外一条皮带继续输送,这也就确保皮带或DCV线发生故障时,不影响值机岛的开放、不影响行李值机过程、不影响行李正常运行过程。

7.1.5.2 直通线应急

为应付DCV故障甚至瘫痪的状况下,在DCV装设备前,设计了直通线,作为备份用途。

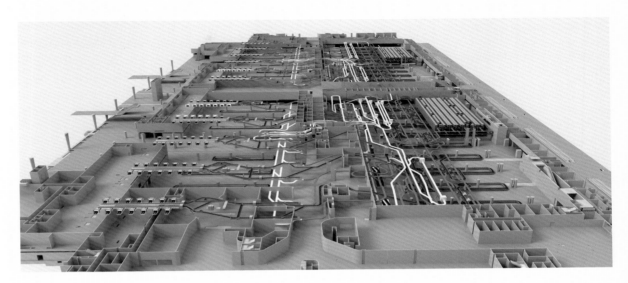

二号航站楼行李系统线路示意
Baggage System Routes in T2

7.1.5.3 DCV系统应急

系统以"独立的模块化"设计,只要维保单位按制造商推荐例行正确地检修DCV系统,DCV发生整套系统全瘫痪/故障的几率非常低。

考虑到DCV分拣系统是作为行李系统的核心机械设备,对于始发、早到和中转行李处理,扮演着非常重要的角色,因此必须周详考虑DCV分拣系统的应急预案。二号航站楼DCV系统的应急预案分几个层次,各层次应急预案具体如下:

1. 每条始发和中转行李的路都由衔接两条主DCV预分拣线,因此可以导入任何一条预分拣线,确保在任何一条预分拣线发生故障时,行李仍可以导入正常运行预分拣线进行分拣;
2. 每条主分拣线都衔接各个装运装置,任何一条主分拣线发生故障时,正常运行的分拣线仍然可以将行李排到任何一个装运装置,不会减少装运装置数量,也不影响行李正常运行;
3. 系统细分为两个不同区域,取分别的电源,确保在A、B其中一个电源中断时不影响行李的正常分拣;
4. 虽然PLC的故障率非常低,每台分拣机还是设置备份PLC,以确保PLC发生故障时,不影响DCV系统正常运行;
5. 行李系统采用RFID行李识别,因此可以充分利用RFID可储存信息的功能作为DCV分拣的应急预案之一;
6. 若是航班信息系统发生故障,无法给行李系统发送所需信息,可通过人工编码站,进行目的地分拣至行李所属的转盘。

以上六层预案作为DCV分拣系统的应急措施,以确保行李系统无论是机械、信息或电源发生故障时,都可以有效迅速地处理行李。

7.1.5.4 中转行李应急

中转系统的输入输送设备都设计在行李分拣大厅的中间部。除无安检机的中转子系统(只供国内转国内的中转行李之外)其余的5个中转子系统都可以输入其他中转行李如国内转国际或国际转国内或国际转国际的中转行李。确保任何一个中转系统故障时,不影响中转行李的输入与运行。

7.1.5.5 早到行李储存系统应急

机械结构方面,早到系统的设计原则和整体系统一致,满足任意一处机械故障均不会影响任意一件行李进出早到系统。

控制方面,除了高层IT提供早到行李的控制信息外,每个早到系统模块将设有本地控制器,而且早到行李的关键信息也将实时存入PLC层:本地控制器实时从高层IT备份本模组内的行李信息,在高层IT出现崩溃无法提供支持时,本地控制器将检查本模组内的早到行李清单,继续按既定的航班时刻释放行李,但不再接受新的早到行李。

在本地控制器崩溃的情况下,高层IT端可启用虚拟EBS控制器,替代本地控制器。已存储的早到行李的航班信息及开放时间也被实时写入PLC层,并不间断更新。出现本地控制器与高层IT均无法支持系统的极端情况,可运用此底层的清单信息,模仿巷道式系统的按航班开放时间优先级释放行李,2小时内清空货架。

7.1 BAGGAGE

7.1.1 General

The baggage system of the Project aims to achieve the maximum transport efficiency, safety and reliability. It has sufficient flexibility and expansion capability to adapt to the future possible demand growth.
Given the target passenger traffic of 45 million in 2020, passenger flows in Terminal 2 (T2) are planned giving consideration to high proportion of transfer passengers to facilitate airline hub operation and passenger trips.

7.1.2 Design Features

The design adopts pallet DCV-aided baggage sorting system, the first application of such technology in China. The system shows significantly improved functionality and technology compared with the dumping tray sorting system of traditional conveyors. Specifically:

1. It adopts DCV-aided trolley automatic sorting system, characterized by fast speed, short time, high efficiency and accuracy. Its DCV automatic addressing function can also improve the operation reliability of the baggage system.
2. It uses efficient and convenient connecting baggage handling system that adds to the hub function of T2.
3. It adopts RFID technology that can improve the baggage tracking rate to around 99.91%. This ensures the offline inspection of the baggage rejected at the security check and customs;
4. It builds a self-service baggage check-in system that can offer better travel experience and higher operation efficiency. Its centralized auto baggage check-in counters, equipped with self-service check-in devices designed as per the size of traditional counters compatible with dual-channel security check machine, will gradually replace the existing traditional check-in counters.
5. It boasts a large-capacity early baggage storage (EBS) system to deliver the baggage of early or long-stay passengers to the baggage handling and storage system via check-in or transfer procedures.
6. It features modular design that reduces the types of spare parts for the baggage system and ensures better expansibility of the DCV system, facilitating future replacement and capacity extension of the system and lowering the risk of construction without suspending air service.
7. It reserves an interface to connect with the baggage system of Terminal 1 (T1), so as to allow for potential connecting baggage handling or integrated operation between T1 and T2.

7.1.3 Baggage Handling Flows

7.1.3.1 Handling Principles
The baggage system consists of departure baggage handling system, arrival baggage handling system, connecting baggage handling system, early baggage handling system, large baggage handling system and VIP baggage handling system.

7.1.3.2 Departure Baggage Handling System
Through the departure baggage handling system, departure baggage is collected at the check-in counter and, after passing the baggage security check system, check-in counter baggage conveyor line, DCV system, EBS system (for early baggage only), manual encoding station and departure shipment carousel, delivered by the baggage van to the aircraft of the designated flight.

7.1.3.3 Arrival Baggage Handling System
Through the arrival baggage handling system, arrival baggage in standard size is delivered by the baggage trailer to the baggage sorting machine room and further to the baggage unloading area on the ground floor, unloaded to the arrival baggage loading-unloading line and then conveyed to the arrival baggage claim carousel via the delivery trench. There are two types of arrival baggage claim carousels defined by length. Each long carousel (about 100 m) corresponds to 2 conveyer lines and each short one (about 70 m) corresponds to 1 conveyor line. The baggage claim hall has 21 standard oblique baggage claim carousels (11 for domestic flights and 10 for international flights).

Arrival baggage in super large size is conveyed to the super large baggage arrival counter in the passenger arrival area via the super large arrival baggage loading-unloading line. Special baggage and baggage that cannot be conveyed by the super large baggage conveyor line (e.g. sailboard) is delivered to the super large baggage arrival counter by hand.

7.1.3.4 Connecting Baggage Handling System
Based on the transfer circulation, the connecting baggage handling system is divided into two systems serving backstage connecting baggage and re-checked connecting baggage respectively.

1. Backstage Connecting Baggage
Baggage claim is unnecessary for passengers of connecting flights, as the baggage can be simply transferred in the backstage baggage sorting machine room. The baggage system of T2 has 6 connecting baggage input routes distributed in the middle of the baggage sorting room, including domestic-international, international-domestic, international-international, and domestic- domestic transfer routes.

The connecting baggage is delivered by the baggage trailer to the transfer line loading-unloading point of the baggage sorting machine room, and is unloaded and input into the connecting baggage input line. Except the input route for domestic-domestic transfer, each connecting baggage input route is connected to one security check machine to check all international flight-related connecting baggage. Rejected suspected baggage is sorted by the DCV system to the corresponding check area for unpacking examination. Baggage passing the security check is sorted to the corresponding carousel. As the main sorting systems in the international and domestic flight areas are already connected through in the design, theoretically, connecting baggage of international-domestic, international-international, and domestic-international connecting flights can all be loaded at any input route of the international or domestic baggage

DCV小车自动分拣行李系统
DCV-aided Trolley Auto Baggage Sorting System

transfer system for transfer.

2. Re-checked Connecting Baggage
As for connecting flights that do not enjoy baggage check-through policy, passengers need to claim their baggage in the arrival area for recheck. To better service these passengers, the baggage system also includes 2 baggage recheck areas at the arrival exit in view of the flow, containing zones for international-domestic transfer, domestic-international transfer, and domestic-domestic transfer. The baggage is handled in the same manner as the departure baggage. Baggage rejected by the CIQ is sorted by the DCV system to the corresponding area for inspection. Baggage passing the inspection is sorted to the corresponding carousel.

7.1.3.5 Large Baggage Handling System
1. Large Departure Baggage
Large departure baggage undergoes the large baggage manual handling flow. Passengers check their baggage at the domestic or international large baggage check-in counter. Baggage rejected by security check is unpacked in situ for examination. Baggage passing security check is delivered by hand via elevator to the baggage sorting machine room for handling.

2. Large Arrival Baggage
Large arrival baggage is delivered by the baggage trailer to the large arrival baggage loading-unloading line for unloading, and then delivered via the large arrival baggage conveyor line to the super large baggage claim hall in the passenger arrival area. Special baggage and baggage that cannot be conveyed by the super large baggage conveyor line (e.g. sailboard) is delivered to the super large baggage arrival counter by hand.

3. Large Connecting Baggage
Large connecting baggage undergoes the large baggage manual handling flow, delivered by the baggage trailer to the security check machine for

large baggage transfer in the baggage sorting machine room for security check, and if accepted, eventually to the large baggage departure flow.

7.1.3.6 Early Baggage Handling System

The domestic area and the international area of T2 are respectively provided with two rack-type EBS systems, each with 2,000 storage spaces, which can be expanded to no less than 6,000 in the long term. Under special circumstances, the domestic area and the international area can share storage resources. Early baggage is stored at the storage spaces inside the rack through random distribution. In addition to early baggage storage, the EBS system also serves for second-tier empty tray caching.

Early baggage may be either departure or connecting baggage. Early departure baggage is delivered by the pre-sorting sub-system to the EBS system; early connecting baggage is delivered to the main sorting system and then sorted to the EBS system. DCV-aided auto rack EBS system is adopted. Each stocker has a minimum handling capacity of 100 pieces per minute.

7.1.3.7 VIP Baggage Handling System

VIP baggage only contribute small load to the entire baggage system and is usually in small amount. The VIP baggage handling system is designed as an independent sub-system without connection to the general baggage system.
VIP departure baggage is checked at the check-in counter of the VIP room, then sent to an independent X-ray machine for security check, and if accepted, eventually delivered by the baggage trailer to the corresponding flight.
VIP arrival baggage is delivered by the baggage trailer directly to the VIP baggage arrival area, and eventually by hand to the VIP lounge.

7.1.3.8 Reserved Room for Baggage System

All interfaces for long-term system expansion is reserved with a whole set or sets of equipment in the same mechanical size of the interfaces. In principle, any whole set or sets of conveyors at any position of the DCV system can be mechanically replaced by 45-degree horizontal shunt and/or 90-degree lateral shunt without any mechanical alteration to the interface upstream and downstream equipment. Possible expansion of the baggage system includes:
1. Additional DCV sorting loop;
2. Additional centralized CT security check area;
3. Expansion of ECS capacity from 2,000 pieces to 3,000 pieces;
4. Expansion of EBS capacity from 4,000 pieces to 6,000 pieces.

7.1.4 Baggage Security Check System

7.1.4.1 Departure and Rechecked Connecting Baggage

To meet the standards of Civil Air Administration and more effectively handle the security check mode of unpacking in the passengers' presence, decentralized counter security check mode is adopted for departure and rechecked connecting baggage. Such mode corresponds to five security check levels. Level 1 involves counter-type single/dual-channel security check machines, which can indicate suspected area in the machine generated images. Level 2 involves security operators who fully study the images generated at Level 1 and send rejected baggage to the unpacking room at the end of the check-in island. Baggage entering the unpacking room generally directly undergoes Level 5 unpacking examination procedure; while baggage containing high-danger prohibited articles undergoes first Level 3 CT examination and then Level 4 manual treatment via ETD equipment.

For baggage security check, connecting baggage is taken to the check-in dual-channel security check machine for security check operators to remotely judge the generated images. Baggage failing the security check are sent to the unpacking room for explosive tracking detection (ETD), followed by Level 5 baggage unpacking examination in the passengers' presence.

7.1.4.2 Backstage Connecting Baggage

Centralized five-level security check mode is adopted for backstage connecting baggage in T2. Level 1 involves double viewing angle high-speed security check machines, which can indicate suspected area in the generated images. Level 2 involves security operators who fully study the images generated at Level 1 and send rejected baggage via BHS to designated inspection station. Baggage entering the inspection station generally directly undergoes Level 5 unpacking examination procedure; while baggage containing high-danger prohibited articles undergoes first Level 3 security check and then Level 4 manual treatment via ETD equipment. The front end of each backstage connecting baggage input line (except domestic-domestic transfer) is connected to a security check machine to examine international flight-related connecting baggage. The baggage is taken to the security check machine for the security check operators to remotely judge the generated images; baggage failing the security check is sorted by the DCV to the corresponding inspection station for Level 5 unpacking examination in the passengers' presence.

7.1.5 Baggage System Emergency Design

7.1.5.1 Check-in Collection Belts

Crossing lines are provided between check-in collection belts. In case any check-in collection belt or DCV line breaks down, the baggage would be conveyed to another belt via horizontal shunt to continue the flow. This guarantees normal operation of the check-in islands and baggage check-in and flow even when the belt or the DCV line is out of service.

7.1.5.2 Through-line

To cope with DCV failure or collapse, through-line is designed as backup before equipment installation of DCV.

7.1.5.3 DCV System

The system adopts independent modular design. As long as the maintenance operator provides regular and correct maintenance recommended by the manufacturer, the probability of collapse/

failure of the DCV system is very low.

The DCV sorting system is the core mechanical equipment of the baggage system, playing a very important role in handling departure, early and connecting baggage. This requires thorough consideration to its emergency plan. The emergency plan for the DCV sorting system of T2 is defined by multiple levels, specifically:

1. Each departure and connecting baggage route is connected to two main DCV pre-sorting lines, and the baggage may enter either pre-sorting line. In case one pre-sorting line breaks down, the other one can still guarantee normal input of baggage;
2. Each main sorting line is connected to various loading devices. In case one main sorting line breaks down, other normally operating sorting lines can still arrange the baggage unto any loading device without reducing the number of loading devices or affecting the normal flow of the baggage;
3. The system is divided into two different zones with respective power sources, so that the interruption of power source in either Zone A or B will not affect normal baggage sorting;
4. Although the probability of a PLC fault is very low, each sorting machine is provided with a backup PLC to avoid affecting the normal operation of the DCV system;
5. RFID baggage identification that allows the function of information storage is adopted as one of the emergency plans for DCV sorting.
6. Where the malfunction the flight information system makes it impossible to send necessary information to the baggage system, manual encoding station may be used to sort the baggage by destination and send them to the corresponding carousel.

The above six-level plans serve as emergency measures for the DCV sorting system, so that the baggage can be effectively and rapidly handled in spite of mechanical, information or power failure of the baggage system.

7.1.5.4 Connecting Baggage

The input and conveying equipment of the transfer system is provided in the middle of the baggage sorting hall. Except for the security check machine-free sub-system (excluding the connecting baggage of domestic-domestic connecting flights), other five transfer sub-systems all support the input of baggage from other connecting flights (such as domestic-international, international-domestic, or international-international ones). In this way, the fault of any transfer system will not affect the input and operation of the connecting baggage.

7.1.5.5 EBS System

The mechanical structure design principles of the EBS system are consistent with those of the whole system, i.e. a mechanical fault at any point will not affect the access of any piece of baggage to/from the system.

In terms of control, the high-level IT provides the control information of early baggage. Besides, each module of the system is provided with a local controller. Key early baggage information is also stored into the PLC level in real time. While the local controller copies the baggage information in the module group from the high-level IT, in case of high-level IT collapse, it would check the early baggage list in the module group and continue to release the baggage by the preset flight time without accepting new early baggage.

In case the local controller collapses, virtual EBS controller may be enabled at the high-level IT end to replace the local controller. The stored flight information and opening time for the early baggage is also written into the PLC level in real time with uninterrupted updating. Under extreme circumstances where neither the local controller nor the high-level IT can support the system, the bottomed list information can be used imitating the alley-type system to release the baggage by the priority of flight opening time and empty the rack within 2 hours.

行李提取大厅
Baggage Claim Hall

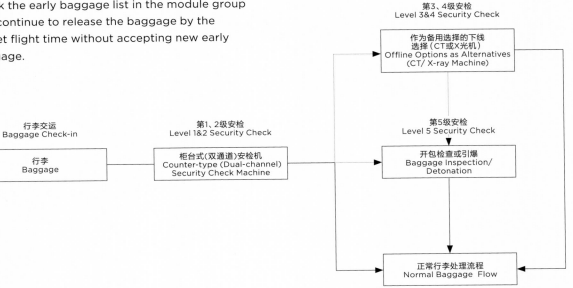

柜台式行李安检流程示意图
Counter-type Baggage Security Check Flow

7.2 结构设计

7.2.1 工程概况

7.2.1.1 工程概况

二号航站楼主楼平面外轮廓尺寸为643m×295m，指廊长度超过1000m。主楼下部有地铁、北进场隧道和城际轨道南北穿过；局部设一层地下室，为设备管廊和行李系统地下机房；地上混凝土结构主体为五层，钢结构网架屋面高度38.1m~44.6m。指廊地上二~四层，地下设备管沟一层，建筑总高约15.07m~26.80m。

交通中心及停车楼位于二号航站楼主楼的南面，作为二号航站楼的配套服务设施。平面外轮廓尺寸为376m×180m。地下二层为人防地下室；地上三层，层高3.75m，建筑高度约14.77m。

7.2.1.2 设计条件

1. 地勘报告概述

项目位于广花凹陷盆地内，区域上地貌单元总体属于冲积阶地地貌，基岩大部分均为第四系土层覆盖。岩土层自上而下分别为填土层Qml、冲积层Qal（中粗砂、黏土、粉细砂、砾砂、粉质黏土）、残积层Qel以及灰岩，其中基岩、不良地质和地下水情况简述如下：

（1）灰岩。岩石裂隙很发育，多为碳质和方解石脉充填。其中夹有溶洞：半充填。主要充填软塑状的黏性土，不均匀发育小石芽，溶蚀发育。

（2）不良地质。不良地质为石灰岩岩溶发育。

航站楼详勘完成的924个钻孔，共48个钻孔揭露到土洞，土洞见洞隙率为5.2%；261个钻孔揭露到溶洞，溶洞见洞隙率28.2%，土洞揭露洞高1.10~18.20米、溶洞揭露洞高0.10~18.1米，主要呈半充填状态，少量无充填、全充填状态，充填物由软塑状态黏性土混约5%~30%的石英质中粗砂组成，局部夹少量灰岩碎石，充填物其强度极低。交通中心区域详勘完成246个钻孔，共7个钻孔揭露到土洞，土洞见洞隙率为2.8%，109个钻孔揭露到溶洞，溶洞钻孔见洞隙率44.3%，线岩溶率21.4%。

（3）地下水。抗浮设防水位取室外地坪标高。勘察场地属Ⅱ类环境，地下水水质在强透水性地层中对混凝土结构具弱腐蚀性，在弱透水性地层中对混凝土结构具微腐蚀性；对钢筋混凝土结构中钢筋具微腐蚀性。

2. 地震安全性评价概述

项目委托广东省地震工程勘测中心进行地震安全性评价，工程区域地壳稳定，近场区内无晚第四纪活动构造，场地未发现断裂通过，场地50年超越概率63%、10%、2%的地面设计峰值加速度分别为25、62、121cm/s²，地震孔中存在砂层，根据计算，均不出现砂土液化。

7.2.1.3 风洞试验

基本风压取50年重现期为0.5kN/m²，地面粗糙度类别为B类。结构体型复杂，对风荷载比较敏感。项目委托广东省建筑科学研究院进行刚性模型测压风洞试验及风致响应和等效静力风荷载研究并做了风环境评估。

7.2.1.4 设计标准

1. 结构设计使用年限。建筑的设计基准期为50年；航站楼的设计使用年限在承载力及正常使用情况下为50年；耐久性下重要构件为100年，次要构件为50年；

2. 建筑安全等级。建筑物安全等级为一级，重要性系数$\gamma_0=1.1$；

3. 抗震设计准则。本工程抗震设防烈度为6°，设计地震分组为第一组，设计基本地震加速度值为0.05g，特征周期0.35s；场地为中软土，场地类别为Ⅱ类场地。按《建筑工程抗震设防分类标准》（GB50223-2008）第5.3节：航站楼主楼及指廊属于重点设防类（乙类）；

4. 地基基础设计等级。基础设计等级为甲级；按《建筑桩基技术规范》（JGJ94-2008）建筑桩基设计等级为甲级；

5. 变形标准。正常使用极限状态下：屋面网架挠度控制为结构空间跨度的1/250；多遇地震作用下，钢结构柱顶侧移控制值为层高H的1/250，混凝土框架柱为1/550；在风荷载作用下，钢结构柱顶侧移控制值为层高H的1/350，混凝土框架柱为1/350；在罕遇地震作用下弹塑性层间位移为层间高度的1/50。

6. 温度荷载。综合考虑：混凝土构件：±15℃；室内钢构件：±25℃；室外钢构件：±35℃。

楼（屋）面活荷载表
Roof Live Load Table

部位 Section	标准值 (kN/m²) Standard Value (kN/m²)	备注 Remarks
楼层典型区域 Typical Floor Area	3.5	候机大厅、走道、贵宾区、办公区等 Waiting hall, walkway, VIP area, office area, etc.
机械、电力、电讯机房、电梯机房 Mechanical, power, telecom and elevator machine rooms	9.0	
储藏室 Storage room	9.0	
行李机房 Baggage sorting machine room	12.0	
行李转盘及通道 Baggage carousel and channel	15.0	
商业区域 Commercial area	5.0	
楼梯间 Staircase	4.0	
不上人屋面（网架上弦）Non-accessible roof (grid upper chord)	0.5	
玻璃纤维顶 Glass fiber roof	0.3	
玻璃纤维顶 Glass fiber roof	0.5	
钢屋盖悬挂活载（网架下弦） Hanging live load of steel roof (grid lower chord)	0.5	不包括检修通道自重及其检修活载 Excluding the dead weight of the maintenance access and its maintenance live load
检修通道检修活载 Live load of maintenance access	2.0	检修通道宽度按1.0m Maintenance access width is 1.0m
非固定隔墙等效活载 Equivalent live load of non-fixed partition wall	4.0	包括无梁支承隔墙 Including beamless partitions
小钢屋等效活载（空调机房） Equivalent live load of steel hut (AC room)	3.0	包含梁、柱、屋面板、管道、吊顶自重及屋面活载，未包含隔墙自重 Including beams, columns, roof slabs, pipes, ceiling dead weight and roof live load, excluding the dead weight of the partition wall
小钢屋等效活载（其他房间） Equivalent live load of steel hut (other room)	2.5	包含梁、柱、屋面板、吊顶自重及屋面活载，未包含隔墙自重 Including beams, columns, roof slabs, ceiling dead weight and roof live load, excluding the dead weight of the partition wall
交通中心停车楼 GTC parking building	4.0	

7.2 STRUCTURE

7.2.1 Project Profile

7.2.1.1 Project Profile
The planar outline dimensions of the main building of T2 are 643 m X 295 m, and the length of pier exceeds 1,000 m. Under the main building are north-south running metro lines, north approach tunnel and intercity railway. One basement floor is provided partially as MEP tunnel and MEP room of baggage system. Five above-grade floors are concrete structure, with the height of steel structure truss roof ranging between 38.1 m to 44.6 m. The piers occupy 2F to 4F while the MEP trenches are provided on B1. The total height of the building is between 15.07 m and 26.80 m.

The GTC and parking building on the south of the main building of T2 serve as the latter's supporting service facilities with a planar outline dimensions of 376mX180m. Two basement floors serve the civil air defense purpose, while three above-grade floors, each in a floor height of 3.75 m, measures a total height of about 14.77 m.

7.2.1.2 Design Conditions
1. Geological Report
Located in Guangzhou-Huadu sag basin, the site features alluvial terraces in terms of the geomorphic units, with the bedrocks mostly covered by the Quaternary soil layers. From top to bottom the rock-soil layers are respectively backfill soil layer Qml, alluvial layer Qal (containing medium-coarse sand, clay, silty-fine sand, gravelly sand, and silty clay), eluvial layer Qel, and limestone, among which bedrock, unfavorable geology and groundwater are briefly described as follows:

(1) Limestone. The rock fissures are well developed, mostly filled with carbonaceous and calcite veins. Karst caves are half filled, mainly with soft plastic cohesive soil, with unevenly developed small clints in corrosion development.
(2) Unfavorable geology is the limestone karst development.

Among 924 boreholes completed in the detailed survey for the site, 48 boreholes expose soil caves with cave fissure rate of 5.2% and cave height between 1.10 m and 18.20 m, and 261 boreholes expose karst caves with cave fissure rate of 28.2% and cave height between 0.10 m and 18.1 m. They are mostly half filled, with a few unfilled and fully filled. The filling, with extremely low strength, is soft plastic cohesive soil mixed with 5% to 30% quartz medium-coarse sand and partially mixed with a small amount of limestone gravel. Among the 246 boreholes completed in the detailed survey for the GTC, 7 boreholes expose soil caves with cave fissure rate of 2.8%, and 109 boreholes expose karst caves with cave fissure rate of 44.3% and linear karst rate of 21.4%.

(3) Groundwater. The anti-uplift water level is subject to the outdoor ground elevation. The surveyed site is a Class-II environment. The groundwater in the highly permeable layer is weakly corrosive to the concrete structure and that in the lowly permeable layer is slightly corrosive to the concrete structure. It is slightly corrosive to the rebar sin the reinforced concrete structure.

2. Seismic Safety Evaluation
The seismic safety evaluation is completed by Guangdong Earthquake Engineering Survey Center. The site has stable earth crust without active late Quaternary structure or fault passing through. With the exceedance probability at 63%,10% and 2% in 50 years, the site has a peak ground accelerations of 25 cm/s^2, 62 cm/s^2 and 121 cm/s^2 respectively. Sand layer is exposed in the seismic boreholes, and no sand liquefaction will occur according to relevant calculation.

7.2.1.3 Wind Tunnel Test
The reference wind pressure is set as 0.5kN/m^2 whose recurrence interval is 50 years, and the ground roughness is Category B. The structure is complex and sensitive to wind load. Guangdong Provincial Academy of Building Research was engaged to conduct the wind tunnel test with rigid model for pressure measurement, the wind-induced response and equivalent static wind load research, as well as the wind environment assessment.

7.2.1.4 Design Criteria
1. Designed service life of structure: with design reference period of 50 years for buildings, the designed service life is 50 years for terminal building given allowable load capacity and normal use, and 100 years and 50 years respectively for important members and secondary members in terms of durability.
2. Building safety grade: Grade I with the importance factor $\gamma O=1.1$.
3. Seismic design criteria. The project is designed as Grade VI in seismic fortification intensity, Group I in seismic group, 0.05g in basic seismic acceleration, and 0.35s in characteristic period. The site with medium-soft soil is a Category II site. The main building and piers of the terminal belong to key fortification class (Class B) according to Section 5.3 of *Standard for Classification of Seismic Protection of Building Constructions (GB50223-2008)*.
4. Designed foundation grade: Grade A for foundation, and Grade A for building pile foundation according to *Technical Code for Building Pile Foundation (JGJ94-2008)*.
5. Deformation criteria. Under normal use limit conditions: the deflection of roof truss is controlled at 1/250 of the structure space span; under the action of frequent earthquakes, the lateral displacement of steel structure column top is controlled at 1/250 of the floor height H, and that of concrete frame column top 1/350; under the action of rare earthquake, the elastic-plastic inter-floor displacement is controlled at 1/50 of the inter-floor height.
6. Temperature load. Comprehensive consideration: concrete members: ±15℃; indoor steel members: ±25℃; outdoor steel members: ±35℃.

7.2.2 结构设计

7.2.2.1 概述

通过设置温度缝（兼防震缝作用）将结构分割成数个较为规则的结构单元：北指廊3个结构单元；东西指廊各5个结构单元；其中主楼首层楼盖由于与基础相连，未设置结构缝，为最大结构单元，最大长度为579m；主楼2层及以上分6个结构单元；考虑旅客自动捷运系统（APM）运行易受到振动的影响，设缝与主楼分开，由于APM部分狭长，APM分为5个结构单元。分缝后主楼上部楼层结构最大长度为216m，指廊的结构最大长度为198m。

柱距主要为9m、18m，最大悬挑跨度为9m。支承混凝土楼盖的柱为钢筋混凝土圆柱；航站楼南侧采用混凝土柱接V字钢柱支承钢结构屋面；内部支承钢屋盖的柱为圆钢管混凝土柱。主楼楼盖采用现浇钢筋混凝土井字梁板体系，指廊楼盖采用现浇钢筋混凝土梁板体系。内部设置较多的连接钢桥，均采用橡胶支座的弱连接方式与主体结构连接。首层行李系统范围采用现浇混凝土预应力空心楼盖。地铁、北进场隧道、城轨范围内的下部轨道结构仅竖向构件与二号航站楼共用。其中城轨下部结构均采用钢管柱，钢管柱不伸出顶板，上部航站楼钢筋混凝土柱与下部连接采用插入钢管内的做法。

屋面均采用正放四角锥网架结构，采用焊接球节点，主体钢结构材料均采用Q345B，网架的上、下表面均为空间曲面。主楼屋盖纵向跨度为54m，45m，54m，横向36m，前端悬挑18m，网架高度为2.5m，沿网架主受力方向设置带肋网架，局部网架总高度为6m并进行立面抽空处理。由于整个钢结构屋盖覆盖范围为东西向约578m，南北向约268m，属于超长结构，考虑温度影响，沿南北向设置两道、东西一道结构缝，将屋面结构划分为6个区。指廊屋盖采用网架高度2.6m。钢屋盖采用钢球铰支座与钢筋混凝土柱顶连接。

7.2.2.2 基础设计

支承上部各楼层结构柱下基础采用端承型冲（钻）孔灌注桩，持力层为微风化灰岩，有Φ800，Φ1200，Φ1400，Φ2200四种直径，单桩承载力特征值3750~260000kN，桩长18m~68m。

地下行李系统和登机桥的基础采用摩擦端承型预应力管桩（PHC500-AB），桩长控制大于18m，为减沉疏桩基础。航站楼无地下室，无整体抗浮问题，但局部有-4.8m~-5.4m的设备和行李管沟，局部考虑水浮力计算，抗浮水位为室外地坪标高，此范围管桩兼作抗拔桩。

场地内石灰岩岩溶发育，要求所有冲孔灌注桩进行超前钻探，按Φ800桩一孔，Φ1000和Φ1200桩二孔，Φ1400桩三孔，Φ2200桩四孔；登机桥的管桩基础按每承台一孔，均匀对称原则布置，超前钻见土洞孔110个，揭露土洞高度1.1~13.7m，见溶洞孔2976个，揭露溶洞高度0.1~26.8m，岩溶发育区域不均衡，个别区域线岩溶率12%，个别区域高达70%。超前钻钻进深度以65m控制，有37个钻孔，孔深达到65m，持力层厚度仍不符合端承要求，设计采用摩擦复核和增大局部位置桩径的处理方式。

针对岩溶地区，开展了钻探结合物探的管波试验，完成了"灌注桩桩基施工期间同步进行的大直径灌注桩控壁岩体完整性探测方法"的发明专利[证书号：2891085]。地下管沟完成后，开始进行工程沉降观测，每施工完一层结构层，监测一次，钢屋面安装就位前和安装完成后各监测一次，以后每三个月进行一次沉降观测，到2016年3月，累计完成16次沉降观测，最大沉降量为3.7mm，沉降速率远小于规范限值要求，沉降均匀，未发现异常情况。

7.2.2.3 混凝土结构设计

二号航站楼按设防烈度7度采取抗震构造措施，大跨框架结构的抗震等级为二级，其中APM部分为单跨框架，抗震等级提高至一级。

1. 大型预应力空心楼盖

首层行李系统区域荷载大，跨度大，比较了普通预应力梁板楼盖、普通双向预应力密肋梁板楼盖，密肋楼盖的经济效益，考虑到首层底板采用梁板式施工不便，特别是梁侧砖模砌筑及回填等工序复杂，影响工期，采用平板的密肋预应力空心楼盖以减少工序。楼盖柱上板带部分采用板厚1200mm预应力空心楼盖，其余位置为700mm厚的空心楼盖，做法如图：密肋空心楼盖断面所示。温度预应力筋在柱上板带布置，每肋采用2束7Φs15.2预应力筋。

针对项目情况，完成了："一种能够减轻自重且具有良好结构刚度的混合型楼盖"的发明专利[证书号：2715387]。

2. 指廊顶层层间位移角的限值

在初步设计阶段，支撑屋盖的柱采用了钢管混凝土柱，顶层柱的层间位移角控制在1/250以内，钢管混凝土柱直径可控制在1.2m以内，钢管柱壁厚25mm，混凝土强度等级为C40，承载力及位移满足规范要求。在施工图设计阶段进行优化设计，将指廊的钢管混凝土柱改为普通混凝土柱。经计算，如按照混凝土框架结构的层间位移角限值1/550控制，改为普通混凝土柱后，柱直径为1.8m，局部2.0m才能满足位移角限值，柱截面尺寸不满足建筑使用及空间效果的要求。考虑航站楼为层高较大的框排架结构，柱顶铰接，排架结构的位移比框架结构大，建筑本身的围护结构多为金属或玻璃幕墙，能承受较大的变形，对二号航站楼是否属超限工程以及其他一些结构设计问题向广东省超限高层建筑抗震设防审查专家委员会提出咨询，对层间位移角限值问题的处理如下：

（1）本工程为框排架结构，建议一、二层框架结构的层间位移角按不大于1/550考虑，顶层排架结构层间位移角宜不大于1/250；

（2）在施工图实施阶段，对西指廊支撑钢屋盖的部分柱采用预应力混凝土柱，经有限元分析表明，预应力混凝土柱的层间位移角放松至1/350时仍基本处于弹性状态，柱受拉损伤情况及裂缝宽度较不施加预应力的普通混凝土柱有明显改善，柱截面尺寸比按1/550的层间位移角限值执行时明显减少（约减少30%），建筑效果和经济效益显著。

3. 锥形钢管柱计算长度系数

二号航站楼主楼钢管柱大部分为变截面钢混凝土柱，最大长度为25.625m。对于典型框架柱的计算长度系数，规范中有比较明确的规定，但对于变截面钢管柱却没有明确的规定，因而对此类锥形钢管混凝土柱计算长度系数根据结构的整体屈曲稳定分析结果确定，按以下三个步骤：

（1）基于第一类稳定原理，对整体结构进行线性屈曲稳定分析，得到整体结构各阶屈曲模态以及屈曲临界荷载系数；

（2）检查各阶屈曲模态形状，确定各钢管混凝土柱发生屈曲失稳时的临界荷载系数，计算出结构整体失稳时屈曲临界荷载N_{cr}；

（3）由欧拉临界荷载公式反算该构件的计算长度系数μ。

结构屈曲与荷载分布模式密切相关。通过结构失稳模态分析，锥形钢管柱的计算长度系数取3.7，两端铰接的人字柱根据弹性屈曲分析，面内面外计算长度系数均取1.1。

4. 大跨度混凝土扁梁与双梁夹钢柱节点

普通混凝土柱的梁柱节点为设置刚性柱帽的组合扁梁做法，本工程框架梁跨度为18米，自重大，在保证受压区和抗剪承载力的前提下采用梁掏空处理减小自重，梁截面形状优化为"π形（跨中）+倒π形（支座）组合扁梁"，梁柱节点采用柱帽刚性节点过渡。钢管混凝土柱部分为减少预应力筋穿孔对钢管柱的削弱，采用双梁夹钢管钢柱的做法，双梁节点区采用环形牛腿+钢牛腿+环梁的构造方式，梁柱节点（柱帽区域）设置抗剪钢牛腿。设计时，通过合理设计"柱-节点-组合扁梁"的承载力，希望引导塑性铰发生在柱帽节点与组合扁梁交界处，通过ABAQUS有限元分析，能满足强柱弱梁、强节点弱构件的抗震构造原则。

π形组合扁梁和双梁夹钢柱的结构形式中，梁和柱在节点处没有对齐，仅通过节点构造完成力的传递。梁端剪力通过柱帽传给柱，特别在双梁夹钢管柱节点处，梁端一部分不平衡弯矩是通过围梁对钢管柱前后产生的压力形成的力偶来平衡，并传递给钢管混凝土柱，其余不平衡弯矩则是通过双梁自身的变形来平衡，节点受力复杂，不能简单地对节点域进行刚性假定。通过采用有限元计算合理评估节点域的弹性刚度，反映梁柱错位的实际情况。通过计算构造和有限元分析，节点承载力满足受力要求。

7.2.2.4 屋面钢结构设计

屋面分区和混凝土分区相对应，考虑多点多向地震计算时将各区组装成整个航站楼计算模型进行分析。以下列举主楼典型二区模型计算结果。

结构自振特性：局部振型较多，采用Ritz向量法对整体计算模型的前90阶模态进行分析，保证振型质量参与系数不小于90%。结构各阶振型主要特点表现为：
1. 结构频谱较为密集。
2. 前若干阶振型主要表现为屋面结构和钢管柱的振动，表明屋盖结构的刚度远小于下部混凝土结构刚度。
3. 第1阶振型为X向平动（南北向），第2、3阶振型为Y向平动（东西向），这与沿长向跨度刚度弱有一定关系。
4. 屋面局部振型主要表现为檐口悬挑端的振动，这与檐口悬挑端的刚度较弱有关。

7.2.2.5 健康监测

项目建立一套全生命周期、实时监测的监测系统，监测包括施工阶段监测和运营阶段健康监测两部分，主要监测内容如下：

1. 施工阶段监测。施工阶段的关键施工节点——钢结构合拢以及卸载时钢结构关键位置的应力、变形、稳定等。
2. 运营阶段监测。运营阶段的环境监测；钢结构整体动力特性；钢结构关键区域的风压影响；钢结构关键构件的应力、变形、稳定等；

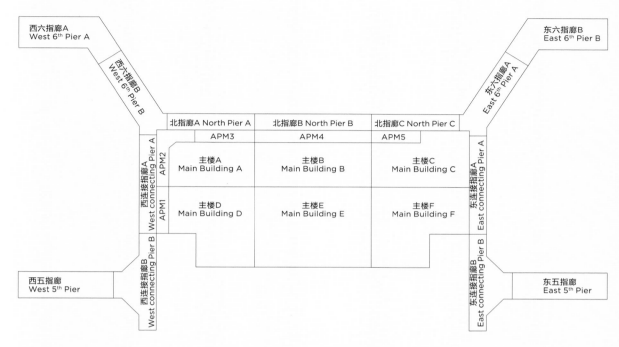

航站楼分区示意图
Terminal Zoning Diagram

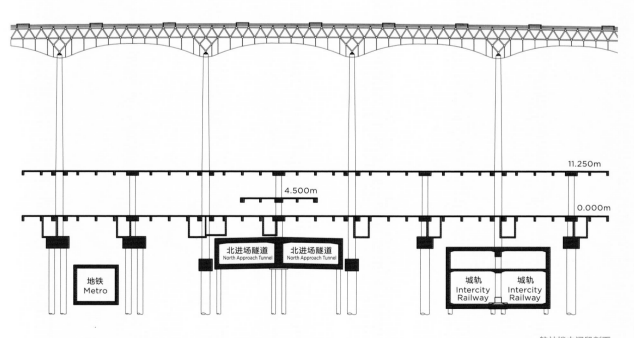

航站楼中间段剖面
Section of Middle Part of T2

7.2.2 Structure Design

7.2.2.1 General

Temperature joints (also used as seismic joints) are provided to divide the structure into several regular structural units, including 3 for the north pier and 5 respectively for the east and west piers. No structural joint is provided on F1 of the main building as its flooring, the largest structural unit with a maximum length of 579m, is connected connected with the foundation. F2 and above floors of the main building are divided into 6 structural units. Given the fact that the APM (automatic passenger mover) operation is susceptible to vibration, joints are provided to separate the system from the main building. The long and narrow APM is divided into five structural units. With the joints provided, the maximum length of the upper floor structure is 216m and that of pier structure is 198m.

The column grid is mainly 9 m and 18 m, with the the maximum cantilever span at 9m. The columns supporting the concrete flooring are round reinforced concrete columns. On the south of the terminal, the steel structure roof is supported by concrete columns that are connected with V-shaped steel columns. The interior columns supporting the steel roof are round steel pipe-reinforced concrete columns. The flooring is the cast-in-place reinforced concrete groined slab system in the main building, and cast-in-place reinforced concrete slab system in piers. For the interiors, many connecting steel bridges are provided to connect to the main building structure via the weak connection of rubber bearings. Within the F1 baggage area, the flooring is cast-in-place concrete pre-stressed hollow flooring. Within metro, north approach tunnel and intercity railway areas, the lower rail structures only share the vertical members with T2. The structures of intercity railway at lower part use steel pipe columns which do not protrude from the top slabs, while the reinforced concrete columns of the terminal structure at upper part are connected with the lower part by inserting into the steel pipes.

The roofing features square pyramid space truss structure with welded spherical joints. The steel structure material of the main building is Q345B, and the upper and lower surfaces of the trusses are spatial curved surfaces. The main building measures the longitudinal spans of 54 m, 45 m and 54 m respectively, and a transverse span of 36 m, with the front cantilever at 18 m and the truss height at 2.5 m. Ribbed trusses are provided along the main stress direction. In certain areas, the total height of the trusses is 6 m with hollowed-out façade. As the entire steel structure roof is an ultra-long structure that spans about 578 m from east to west and 268 m from north to south, two structural joints are provided in north-south direction and one in east-west direction to divide the roof structure into six zones to address the temperature impacts. The pier roof features the truss height of 2.6 m, and the steel roof is connected with the reinforced concrete columns via steel spherical hinge bearings.

7.2.2.2 Foundation Design

The foundation under the structure columns supporting the upper floors uses end-bearing punched (bored) piles with four diameters, respectively φ800, φ1200, φ1400 and φ2200, with the bearing strata being the weakly-weathered limestone. The bearing capacity of a single pile is 3,75 kN to 260,000kN, and the pile length is 18m to 68m.

The foundations of the underground baggage system and the boarding bridges use friction end-bearing pre-stressed pipe piles (PHC500-AB) with length controlled over 18m, which are settlement reducing pile foundations. The terminal has no basement hence is free from anti-uplift problem generally. Yet there are some MEP and baggage trenches of -4.8m to -5.4m deep in some locations. According to the calculation of water buoyancy in those locations, the anti-uplift water level is subject to the outdoor ground elevation, and the pipe piles within the scope are also used as uplift piles.

The site, with developed limestone karst, requires prior boring for all cast-in-place piles as required: one borehole for a φ800 pile, two respectively for a φ1000 pile and a φ1200 pile, three for a φ1400 pile, and four for a φ2200 pile. For the pipe pile foundation of boarding bridge, one borehole is for a pile cap, and all boreholes are distributed evenly and symmetrically. The prior boring exposes 110 soil caves with height of 1.1 m to 13.7 m, and expose 2,976 karst caves with height of 0.1 m to 26.8 m. The karst development areas are unevenly developed, and the linear karst rate is 12% in some areas and could be as high as 70% at several locations. The prior boring depth is controlled at 65 m, and 37 boreholes are

Cross Section Diagram of Hollow-Ribbed Floor

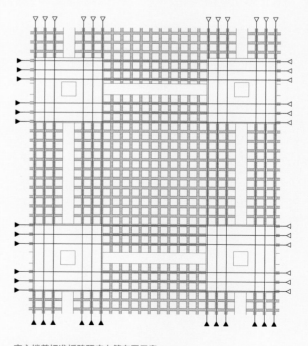

Schematic Layout of Hollow Floor Typical Slab Spanning Pre-stressed Tendons

required with depth of 65 m. The thickness of bearing strata still does not meet the requirements of end bearing, so friction recheck and increased pile diameter in some places are provided.

For the karst areas, the tube wave test of both boring and geophysical exploration is conducted, and invention patent known as Method for Detecting the Integrity of Rock Mass with Wall Controlled by Large-diameter Cast-in-place Pile in synchronization with Construction of Cast-in-place Pile Foundation" has been completed [certificate No.: 2891085]. The settlement observation starts upon the completion of the underground pipe trenches. Each structural floor is monitored once the construction is done. The steel roof is monitored once respectively before and after the installation, which is then followed by a settlement observations every three months. By March 2016, 16 settlement observations were conducted, showing a maximum settlement of 3.7 mm, which was much lower than the limit set forth by relevant codes. Besides, the settlement has been uniform and no abnormal condition was found.

7.2.2.3 Concrete Structure Design

For T2, the seismic structure measures are based on Fortification Intensity Level VII, and the seismic grade of large-span frame structures is Level II, except that the seismic grade of APM structure is raised to level 1 because it is a single-span frame.

1. Large Pre-stressed Hollow Floor

For baggage area on F1, which features considerable load and span, various options are compared in terms of cost-effectiveness, including the ordinary pre-stressed beam-slab floor, ordinary two-way pre-stressed multi-ribbed beam-slab floor, and multi-ribbed floor. In view of the inconvenience in beam-slab construction for F1 base slabs, in particular the impact of the complex procedures of beam-side brickwork and backfilling on construction period, the flat multi-ribbed pre-stressed hollow floor is adopted to reduce the working procedures. For slab strip on

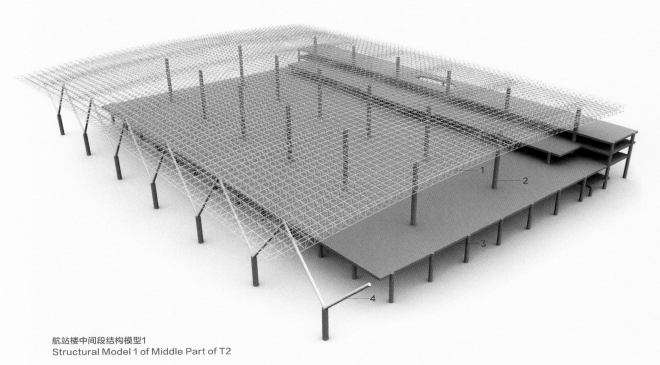

航站楼中间段结构模型1
Structural Model 1 of Middle Part of T2

航站楼中间段结构模型2
Structural Model 2 of Middle Part of T2

1. 带肋钢网架
2. 钢管混凝土柱
3. 下部钢筋混凝土框架楼层
4. Y型柱（下部混凝土柱+V形钢管柱）

1. Ribbed Steel Grid
2. Concrete Filled Tubular Steel Column
3. Lower Floor of Reinforced Concrete Frame
4. Y-shape Column (Lower Part Concrete Column + V-shape Tubular Steel Column)

the floor columns, 1,200 mm thick pre-stressed hollow floor is used; and for other floor areas, the hollow floor with thickness of 700 mm is used. See Figure: Section of Multi-ribbed Hollow Floor for the construction method. The temperature pre-stressed tendons are provided on the slab strips on the columns and 2 bundles of 7Φs15.2 pre-stressed tendons are used for each rib.

To address the the project, an invention patent known as "a hybrid floor with reduced dead weight and good structural stiffness" has been completed [Certificate No. 2715387].

2. Inter-floor Displacement Angle Limit of the Top Floor of Pier

In the DD stage, the columns supporting the roof are steel pipe-reinforced concrete columns. The inter-floor displacement angle of the columns of the top floor is controlled within 1/250, with the diameter of steel pipe-reinforced concrete columns within 1.2 m, the wall thickness of the steel pipe columns at 25 mm, the concrete strength grade at C40, and the bearing capacity and displacement up to the requirements of the relevant codes. During design refinement at CD stage, the steel pipe-reinforced concrete columns are changed into ordinary concrete columns. As a result, only when the column diameter is 1.8 m generally and even 2.0 m in some places can the displacement angle meet the limit requirement, but the section size of columns fails to meet the requirements of building use and spatial effect. Given that the terminal is a frame-bent structure with large floor height and the column tops are hinged, the displacement of the bent structure is larger than that of the frame structure; meanwhile, the building envelop is mostly metallic or glass curtain wall which can withstand great deformation. Regarding whether Terminal 2 is an over-limit project and other structure design issues, the Expert Committee on Seismic Fortification Review of Over-limit High-rise Buildings of Guangdong Province was consulted to address the inter-floor displacement angle limit as follows:

(1) The project is a frame-bent structure. It is suggested that the inter-floor displacement angle of F1' and F2's frame structures be no more than 1/550, and that of top floor's bent structure no more than 1/250;

(2) In the CD implementation stage, pre-stressed concrete columns are used as to support some of the steel roof of the west pier. The finite element analysis shows that the pre-stressed concrete columns are still basically in elastic state when the inter-floor displacement is relaxed to 1/350; the tensile damage and the crack width of the columns are significantly improved in comparison with the ordinary concrete columns without pre-stressing; the column section size is greatly reduced (by about 30%) in comparison with that when inter-floor displacement angle limit is 1/550; and the architectural effect and economic benefits are significant.

3. Coefficient for Effective Length of Tapered Steel Pile Columns

Most of the steel pipe columns used in the main building of T2 are variable-section steel pipe-reinforced concrete columns, with maximum length of 25.625 m. For the coefficient for effective length of typical frame columns, there are clear provisions in relevant codes; but for that of variable-section steel pipe columns, no clear provisions are available. Therefore, the coefficient for effective length of such tapered steel pipe-reinforced concrete columns is subject to the overall buckling stability analysis results of the structure following the three steps below:

(1) Based on the first-category stability principle, the linear buckling stability analysis of the whole structure is carried out to obtain the buckling modes of various orders and critical buckling load coefficients of the whole structure;

(2) The buckling mode and shapes of various orders are checked to determine the critical load coefficient of each steel pipe-reinforced concrete

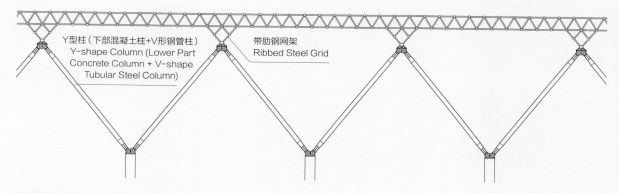

带肋钢网架及Y型柱（下部砼柱+V形钢管柱）
Ribbed Steel Grid and Y- shape Column (Lower Part Concrete Column + V-Shape Tubular Steel Column)

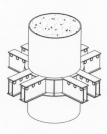

双梁有限元环板与牛腿模型
Double-Beam Finite Element Ring Plate and Bracket Model

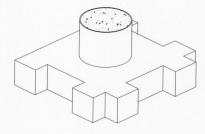

双梁有限元节点模型
Double-Beam Finite Element Node Model

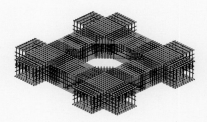

双梁有限元节点区钢筋分布模型
Model of Rebar Distribution in Double-Beam Finite Element Node Area

column when buckling instability occurs, and calculate the critical buckling load Ncr when the whole structure is unstable;
(3) The coefficient Q for effective length of the member is calculated by Euler critical load formula.

Structural buckling is closely related to load distribution pattern. Through structural instability modal analysis, the coefficient for effective length of tapered steel pipe columns is determined as 3.7, and for the A-shaped columns hinged at both ends, through elastic buckling analysis, the coefficient for effective length of such columns inside and outside the surface is determined as 1.1.

4. Joint between Large-span Concrete Flat Beam and Double-beam-clamped Steel Column
For the beam-column joints of ordinary concrete columns, combined flat beams with rigid column caps are used. In the project, the span of the frame beam is 18 m with considerable dead weight. To reduce the dead weight, the beams are hollowed out while ensuring pressure zone and shear bearing capacity. The beam section is optimized as "Ⓟ-shaped (mid-span) + inverted Ⓟ-shaped (bearing) flat beams", while the column cap rigid joints are used as beam-column joints for transition. To mitigate the weakening of the steel pipe-reinforced concrete column by the perforation of the pre-stressed tendons, the double-beam-clamped steel-pipe steel columns are applied. The structure of ring bracket + steel bracket + ring beam is adopted for the double-beam joint parts, while the shear steel brackets are provided for the beam-column joints (column cap parts), as shown in the figure. The bearing capacity of "column-joint-combined flat beam" is reasonably designed for the purpose of limiting the plastic hinge outside the boundary between the column joint and the combined flat beam. As per ABAQUS definite element analysis, the seismic structure principle of strong column with week beam and strong joint with weak member is satisfied.

In the structural form of Ⓟ-shaped combined flat beam and double-beam-clamped steel column, the beam and column are not alighted at the joint, so the force is transmitted only through joint structure. The beam end shear force is transmitted to column through column cap; especially at the double-beam-clamped steel pipe column joint, a part of the unbalanced bending moment at the beam end is balanced by the force couple formed by the pressure generated by the wales on the front and back of the steel pipe column, and transmitted to the steel pipe-reinforced concrete column; the rest of the unbalanced bending moment is balanced by the deformation of the double beams themselves, rendering the joint subject to complex forces, so it is impossible to simply make an assumption of the stiffness of the joint. The finite element calculation is used to reasonably evaluate the elastic stiffness of the joint field to reflect the actual situation of the beam-column misalignment. The structure calculation and finite element analysis show that the bearing capacity of the joint meets the stress requirement.

7.2.2.4 Roof Steel Structure Design
The roof zoning corresponds to the concrete zoning. In view of multi-point and multi-direction seismic calculation, the zones are simulated as the entire terminal calculation model for analysis. The calculation results of the Typical Zone 2 of the main building are listed below.

Self-vibration characteristics of structure: there are many local vibration modes, so Ritz vector method is used to analyze the first 90 modes of the overall calculation model to ensure the modal participating mass ratio is no less than 90%. The main characteristics of each vibration mode are:

1. The structure spectrum is relatively dense.
2. The first several modes are mainly characterized by the vibration of roof structure and steel pipe column, indicating that the stiffness of roof structure is much lower than that of the lower concrete structure.
3. The first-order vibration mode moves horizontally along X-direction (i.e., north-south direction), and the second- and third-order vibration modes move horizontally along Y-direction (i.e., west-east direction), which has a certain relationship with the weak stiffness along the long span.
4. The local vibration mode of roof is typified by the vibration of the cantilever end of cornice, which is related to the weak stiffness of the cantilever end of cornice.

7.2.2.5 Health Monitoring
For the project, a monitoring system is developed for full life circle and real-time monitoring, including two parts: monitoring in the construction stage and health monitoring in the operation phase. The main contents under monitoring are as follows:

1. Monitoring in the construction stage: the key construction milestones – stress, deformation, stability, etc. at the key positions of the steel structure when the steel structure is closed and unloaded.
2. Monitoring in the operation stage: environmental monitoring; overall dynamic characteristics of steel structure; wind pressure influence at key fields of steel structure; stress, deformation, stability, etc. of key members of steel structure.

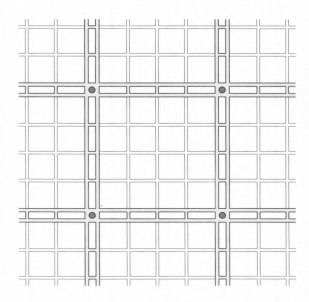

钢管柱双梁标准平面单元
Steel Tubular Column Double-Beam Typical Planar Unit

7.3 机电设计

7.3.1 机电设计亮点

1. 应用了大量的新产品、新技术,为建设平安机场、绿色机场、智慧机场提供有力的技术保障。
2. 深度调研并结合管理与运行需求,针对性设置机电系统,确保航站楼建成后就能用、好用、好管。
3. 在公共空间研究应用大功率LED灯与智能调光控制技术,实现"营造良好的灯光环境,尽可能地降低照明功耗,最大限度地延长灯具寿命"的目的,带来良好的灯光环境,长远的经济效益,以及绿色环保的示范作用。
4. 研究利用二维码技术,创造性将其与建筑智能化系统相结合,通过把机电设备及系统数据的二维码化设计一体化管理平台,实现各机电系统信息共享与资源整合,并为物联网技术应用打下扎实的软硬件基础。
5. 通过建筑智能化系统营造良好舒适光环境、空气环境,为旅客和后勤保障营造舒适的空气环境。
6. 采用大温差7/15℃供回水空调冷冻水系统、高效设备与变频控制、大小机组配合等措施,确保系统安全可靠、节材节能、方便调节和管理。
7. 公共卫生间均设计了空调系统和机械排风系统,且机械排风系统设有上下排风口,提高了环境的舒适性。
8. 全年24小时不间断空调应采用独立的多联机空调系统。
9. 所有制冷主机全部采用变频主机,保证了空调系统的高效。
10. 空调系统设计了初效、中效、静电除尘三级过滤系统。提高了室内人员的舒适度。
11. 行李提取大厅的空调排风到未设空调的行李库,既提高了行李库的舒适性,又减少了行李库的排风量,节约了能耗。

7.3.2 给排水设计

7.3.2.1 概述

二号航站楼各系统在总体延续一号航站楼设计思路基础上,根据现行规范要求进行全面提升。新材料、新技术在设计中得到了广泛的应用。给水系统竖向分为两区,四、五层采用无负压供水系统。排水系统增设化粪池、一体化油水分离装置、一体化污水提升装置等。增设雨水回收利用系统及雨水调蓄系统,减少了雨水排放对周边水体的压力。消防方面增加大空间智能型主动喷水灭火系统对大空间区域全覆盖保护;设置高压细水雾系统对地下管廊电舱的保护。

7.3.2.2 给排水各系统设计特点

1. 室外给水

由广州市江村水厂提供给水水源,经两根DN800的场外输水管线接到机场水表房并最终接入场内供水站。1/3负荷在北区,2/3负荷在南区,南供水站已建两座4000m³水池。北供水站已建两座2000m³水池,预留一座4000m³水池。机场给水管网平面成环状布置,并设置分段分区的检修阀。机场的生活、生产用水与消防用水合用一条管线。机场内给水管网布置满足一期和远期发展的需要。每个配水站的出水总管和场内给水干管可满足远期发展的需求。场内给水干管供水压力0.40MPa,可满足三层供水压力要求。

2. 室外排水

场内采用雨、污水分流制。污水系统为一个独立的管网系统,场内的生活污水和生产废水最终排至机场污水处理厂。现有机场内污水排水管网满足机场本期规模中的二号航站楼及各个分区的建构筑物的排水量要求,并同时考虑了机场近期发展污水量增加的因素。

室外雨水分空侧和陆侧两部分,空侧雨水排到飞行区雨水系统。陆侧部分,目前1号雨水泵站排水能力及空侧排水能力也不足以承接新建二号航站楼陆侧雨水量,须考虑雨水调蓄方案,雨水调蓄池按20年一遇暴雨强度计算,有效容积为2.6万m³,通过进出雨水渠箱及现状1号雨水调蓄池及新建调蓄池水位雍高等方式,可满足50年一遇暴雨强度,保证机场路面无积水。雨v水调蓄池设置在二号航站楼西南侧南往南高架桥西侧转弯位,高架桥下的绿化带内,采用全地埋式,地面仅保留检修人孔及水泵配电柜等。

3. 室内给水

以航站区市政给水管网为生活给水水源。三层及以下各层

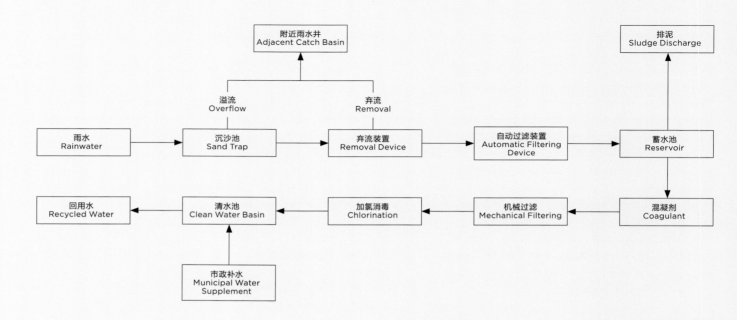

雨水利用流程图
RAINWATER Utilization Flow

由市政管网直接供水，四层以上由无负压供水设备供水，充分利用室外管网压力。生活泵房设于航站楼主楼首层东南角，设无负压供水设备一套。室内给水管网成环布置，环网上设有切换阀门，任一路供水故障或者任一区域故障可通过阀门切换及时检修。按不同使用用途、不同计费单位、区域分区等设置水表，除给水引入总管设置总水表外，公共卫生间、空调机房、厨房用水、商业用水接入端等均设置分区计量水表。所有水表均自带远传通信功能，通讯协议采用M-bus规约，管理中心可随时掌握各处的用水情况。

计时旅馆、头等舱及商务舱区域采用太阳能集中式全循环热水供应系统；贵宾区域及母婴间采用局部热水系统，就近需用热水卫生间设一组容积式电热水器供应热水。

4. 直饮水

直饮水的供应采用各饮水点设置终端过滤饮水机方式。饮水机水源为自来水，净水流量0.3L/min，储水容量：冷水3.9L，热水23L。

5. 室内排水

（1）室内生活排水采用污废分流排放，室外设化粪池、厨房设隔油器。生活污水经化粪池处理、厨房含油废水经隔油器处理后与生活杂排水汇合接入航站区污水管网，然后排往机场污水处理站。

（2）航站楼中部排水点远离室外，最远端超过150m，采用一体化污水提升装置，解决中部排水点距离室外太远的问题。其中二层及以上采用地上安装式污水提升装置，配备带切割装置潜污泵两台、自动耦合装置、控制箱，有效容积1000L。首层中部卫生间设置地埋式污水提升装置，污水提升装置设置于首层专用的设备坑中，配备带切割装置潜污泵两台、自动耦合装置及控制箱，有效容积250L或400L。

（3）地下室及设备管廊设集水坑，由潜水泵将坑内积水抽出室外。急救站、急救室及一层检疫区所排的含有病菌的污水经消毒处理后再排到室外管网。

（4）排水系统分组设置汇合通气管，由屋檐下侧墙通往室外，管口处设百叶窗处理并加设防虫网。

6. 屋面雨水

（1）设计参数。屋面采用虹吸雨水排放系统，金属屋面暴雨重现期取20年，雨水系统与溢流设施的总排水能力按不低于50年重现期的雨水量计算；主楼北侧混凝土屋面设计重现期取50年，雨水系统与溢流系统的总排水能力按不低于100年重现期的雨水量计算。降雨历时均按5分钟计算，屋面径流系数为1。

（2）系统设置。屋面总汇水面积达273200m²，根据屋面造型划分不同的排水区域，其中最大排水区域面积为22500m²。主楼南北侧天沟为坡天沟，坡度从3%～10%变化，东西侧天沟为平天沟。屋面板为直立锁边金属屋面，雨水通过直立锁边高出的侧边，形成小天沟，引导水流流向标高较低的天沟内。雨水斗的布置原则为：平天沟分散布置，坡天沟集中布置。在天沟内每隔一定距离设集水槽，雨水斗设于槽内，使坡天沟可分段截流，避免了天沟最低点水量过大。天沟内设有溢流口，大大增加系统的安全性。主楼北侧约31500m²的屋面花园及设备区域为混凝土屋面，设置内天沟。

7. 雨水回收利用

（1）系统规模。本工程收集部分屋面雨水用作室外绿化、幕墙及道路冲洗。所收集雨水集中在主航站楼屋面南边，总收集面积为约46000m²。结合回用水实际使用量设计2个雨水收集池，设置于主楼首层室外东、西两侧绿化带内，容积均为800m³。

（2）回用水系统。当雨水利用系统回用水箱水位达到低水位时，考虑补充市政水源进雨水清水池。补水时，控制补水至雨水正常蓄水水位的大约2/3处，留出空间以备降雨的收集。二号航站楼东、西边各设置一套雨水回用系统，采用变频供水设备供水。

（3）系统控制。具备自动控制、远程控制、就地手动控制。泵房、楼梯口集水坑及雨水收集池和雨水清水池溢流报警信号引至主楼设备管理中心及TOC。对常用控制指标（水量、主要水位）实现现场监测，补水由水池水位自动控制。

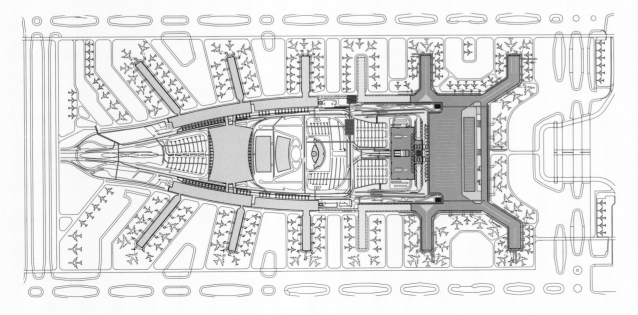

航站区雨水总图
Rainwater Master Plan of Terminal Area

— 原一号航站楼雨水沟（渠）
Original Rainwater Ditches of T1

— 新增二号航站楼雨水沟（渠）
New Rainwater Ditches of T2

■ 原有雨水调蓄池
Original Rainwater Detention Pool

■ 新增雨水调蓄池
New Rainwater Detention Pool

■ 新增雨水收集池
New Rainwater Catchment

7.3 MEP

7.3.1 Design Highlights

1. A great number of new products and new technologies are applied to provide solid technical support for the construction of a safe, green and intelligent airport.
2. Based on in-depth research and in consideration of the management and operation requirements, the MEP systems are planned to ensure easy use, operation and management of T2 upon its completion.
3. High-power LED lamps and intelligent dimming control technology are applied in public spaces to "create a favorable lighting environment, minimize lighting power consumption and maximize the life of lamps". The goal is to achieve a favorable lighting environment, long-term economic benefits, and a demonstration effect of environmental protection.
4. QR code technology is creatively incorporated into the building intelligent systems. With an integrated management platform for QR code design of MEP equipment and system data, information sharing and resource integration of various MEP systems can be realized and a solid foundation of software and hardware provided for the application of IoT technology.
5. The intelligent building system helps create favorable and comfortable daylight and air environment for passengers and BOH staff.
6. Measures such as AC chilled water supply and return system with large temperature difference of 7/15℃, high-efficiency equipment and frequency conversion control, and combination of large and small units are used to ensure the safety and reliability of the systems, cut material and energy consumption, and facilitate system adjustment and management.
7. AC system and mechanical exhaust system are designed in all public washrooms. The mechanical exhaust system is equipped with upper and lower exhaust outlets to improve the comfort level of the environment.
8. An independent multi-split AC system is used for 24-hour uninterrupted AC all year round.
9. All refrigerators are of frequency conversion type to ensure the high efficiency of the AC system.
10. For the AC system, a three-stage filtration system of primary filter, medium filter and electrostatic precipitator is designed to improve the interior comfort level.
11. Air of the AC system in the baggage claim hall is exhausted to the AC-free baggage storage to improve the latter's comfort level and reduce its exhaust air volume and energy consumption.

7.3.2 Plumbing Design

7.3.2.1 General

Continuing the design concept of T1, the systems of T2 are comprehensively upgraded according to applicable codes and specifications. New materials and technologies are extensively applied in the design. The water supply system is vertically divided into two zones, with non-negative pressure water supply system adopted for F4 & F5. Septic tanks, integrated oil-water separation devices and integrated sewage lifting devices are added to the drainage system. Rainwater recycling and storage systems are added to reduce the pressure of rainwater discharge on surrounding water bodies. For fire protection, the design provides intelligent active sprinkling system for large spaces for full-coverage protection of large-space areas and high-pressure water mist system for the protection of electrical cabins in utility tunnels.

7.3.2.2 Design Highlights

1. Exterior Water Supply

The water supply source provided by Guangzhou Jiangcun Waterworks is connected through two off-site water supply pipelines of DN800 to the water meter room in the airport and eventually the on-site water supply station. One third of the loads are in the north area and two thirds in the south area. Two 4,000m³ water tanks are provided in the south water supply station, while two 2,000m³ water tanks are provided and one 4,000m³ water tank is planned in the north station. The water supply pipeline network of the airport displays a circular layout, with maintenance valves provided by segments and zones. The domestic water, production water and fire water share one same pipeline in the airport. The layout of water supply pipeline network can accommodate the needs of both Phase I and long-term development. The water outlet main of each water distribution station and the trunk feed system in the site can meet the needs for long-term development. The water supply pressure of the trunk feed system in the site is 0.40MPa and can supply water for three floors.

2. Exterior Water Drainage

Separate pipes are adopted for rainwater and sewage drainage. The sewage system is an independent pipe network system that discharges the domestic sewage and production wastewater eventually to the airport sewage treatment plant. The existing sewage drainage pipe network of the airport meets the drainage needs of T2 and the buildings and structures in various zones in this phase and takes into account the sewage increase in near-future development.

Exterior rainwater drainage system covers both landside and airside rainwater. Airside rainwater is drained to the rainwater system in the movement area. For landside rainwater, the current drainage capacity of the No.1 Rainwater Pumping Station and the airside drainage capacity are not sufficient to handle the landside rainwater of the newly-built T2. Therefore, a rainwater storage scheme is proposed. The rainwater detention tank is planned with an effective volume of 26,000m³ based on 20-year rainwater recurrence interval. By rising the water level of the inflow and outflow rainwater sewer tanks, the current No.1 Rainwater Detention Tank and the newly-built water storage tank, the system can accommodate rainstorm with a 50-year recurrence interval and ensure the road surface free of standing water. The rainwater detention tank is located at the turning point on the west of the south-to-south viaduct to the southwest of T2. It is completely buried in the green belt under the viaduct with only the maintenance manhole and water pump power distribution cabinet exposed on the ground.

3. Interior Water Supply

The municipal water supply network in the terminal area is taken as the domestic water supply source. Water supply to F3 and lower floors is directly led from the municipal pipe network, while that to F4 and higher floors is achieved by means of non-negative pressure water supply equipment making full use of the exterior pipe network pressure. The domestic water pump room is located at the southeast corner on F1 of the main building of T2, in which a set of non-negative pressure water supply equipment is installed. The interior water supply network is arranged annularly, with switching valves installed to ensure timely repair via valve switching in case of water supply failure of any pipeline or in any zone. In addition to general water meter in the water main, water meters are also provided separately for public washrooms, AC rooms, kitchens, commercial water access terminals, etc. based on different uses, charging units and zones. All water meters are equipped with remote communication function based on M-bus protocol, so that the management center can have access to the water consumption data at any place and time.

Centralized and full-circulating solar water heating system is provided for hourly charged hotels, first class and business class lounges, while local water heating system is provided for VIP area and baby care room, with a set of volumetric electric water heaters provided for nearby washrooms where hot water is needed.

4. Drinkable Water

For supply of drinkable water, dead-end filtration drinking fountains are provided at each drinking point. The water source of the drinking fountain is tap water with a purified water flow rate of 0.3L/min and the water storage capacity of 3.9L for cold water and 23L for hot water.

5. Interior Water Drainage

(1) For interior domestic drainage, sewage and wastewater is discharged separately, with septic tank provided outdoors and oil separator in the kitchen. Domestic sewage treated in the septic tank and kitchen oily wastewater treated by the oil separator is drained into the sewage pipe network in the terminal area, and then to the airport sewage treatment plant.

(2) The central drainage point of the terminal is far away from the outside with the farthest end over 150m. An integrated sewage lifting device is adopted to solve this problem. F2 and higher floors are provided with ground type sewage lifting devices, equipped with two submersible sewage pumps with cutting devices, automatic couplers and control box, totaling an effective volume of 1,000L. The washroom in the middle of F1 is provided with buried sewage lifting device installed in a dedicated MEP pit on F1, equipped with two submersible sewage pumps with cutting device, automatic couplers and control box, totaling an effective volume of 250L or 400L.

(3) Sump pits are provided in the basement and utility tunnel, and the accumulated water in the pits is pumped to the outside by submersible pumps. Sewage containing bacteria discharged from the first aid station, the first aid room and the quarantine area on F1 is disinfected and then discharged to the exterior pipe network.

(4) In the drainage system, vent headers in groups leading from the side wall under the eave to the outside are provided. Louvers are installed at the pipe openings with flynet also provided.

6. Roof Rainwater

(1) Design parameters. Siphon type rainwater drainage system is used on the roof. The metal roof is designed based on the rainstorm reoccurrence period of 20 years, and the total drainage capacity of the rainwater system and overflow facilities is calculated based on a minimum rainwater runoff of 50-year reoccurrence period. The rainwater reoccurrence period adopted for the concrete roof in the north of the main building is 50 years, and the total drainage capacity of the rainwater system and overflow system is calculated based on a minimum rainwater runoff of 100-year reoccurrence period. The rainfall duration is assumed to be 5 minutes, with roof runoff coefficient at 1.

(2) System setting. The total catchment area of the roof is 273,200m^2. It is divided into different drainage areas in view of the roof shapes, with the largest drainage area at 22,500 m^2. The gutters on the south and north sides of the main building are sloped with the gradient varying from 3% to 10%, and the gutters on east and west are horizontal. On the vertical edged metal roof, rainwater flows through the vertical edges that form small gutters to the gutters at lower levels. Flat gutters are decentralized while sloped ones are centralized. Catch tank containing roof drains are provided at certain intervals in the gutter, so that the sloped gutters may cut off the rainwater by section and prevent excessive flow at the lowest point of the gutter. Overflow ports are provided in the gutter, greatly improving the safety of the system. Inner gutters are provided in the concrete roof of the approx. 31,500m^2 roof garden and MEP area on the north side of the main building.

7. Rainwater Recycling

(1) System capacity. Part of the harvested roof rainwater is used for outdoor greening irrigation, curtain wall and road flushing. Rainwater harvesting takes place mainly on the south roof of the main building with a total harvesting area of about 46,000m^2. In view of actual needs, two rainwater harvesting pools are provided in the green belts on the east and west of F1 of the main building, each with a capacity of 800m^3.

(2) Water recycling system. When water in the recycled water tank of the rainwater reuse system drops to the lower level, municipal water would be supplied to the clean rainwater tank until reaching about 2/3 of the normal rainwater storage level, with space reserved for rainwater harvesting. On the east and west of T2, one rainwater recycling system is planned respectively, equipped with frequency-conversion water supply equipment.

(3) System control. Automatic control, remote control and local manual control are all allowed. The overflow alarm signals from the pump room, stairway sump pits, rainwater collection tank and clean rainwater tank are transmitted to the building services management center and TOC of the main building. Regular control indexes (water volume and main water level) are monitored in the field and water replenishment is automatically controlled based on the water level of the tank.

8. 卫生间给排水

卫生间分区域供水，确保排水系统安全；采用支状供水与环状供水相结合，提高供水稳定性和使用舒适性；在每排厕格背墙一侧设置检修通道，方便施工和维护。

9. 消火栓系统

（1）室外消火栓系统。室外消防系统分空侧和陆侧两部分。空侧室外消防系统与飞行区消防系统共用一套管网；陆侧室外消防系统从航站区DN500供水管网接出两根DN300管呈环状布置成室外消防管网。

（2）室内消火栓系统。二号航站楼室内外消防系统分开设置，室内消防系统（包括室内消火栓系统、自动喷水灭火系统、水幕分隔及保护系统）共用一套加压管网。

10. 自动喷水灭火系统

（1）除高大空间、楼梯间、小于5m²的卫生间等不易引起大火的房间及不能用水扑救灭火的部位外，均设置自动喷水灭火系统。一般区域按中危险Ⅰ级考虑，交通中心停车楼按中危险Ⅱ级考虑。

（2）与室内消火栓系统共用加压泵组及加压管网。航站楼内按区域分别设置若干个报警阀间，每个报警阀间由消防加压管网引入两条给水管连接报警阀组，每条引入管处设置检修阀门、电动闸阀和止回阀，每层及每个消防分区均设水流指示器，水流指示器信号在消防中心显示。按区域分别设置消防水泵接合器。

（3）喷头的选择和布置：采用快速响应喷头。有天花吊顶的下向型喷头，采用装饰型隐蔽喷头；高度大于800的天花内设上向型喷头；厨房喷头动作温度93℃，天花内动作温度79℃，其余喷头动作温度68℃。金属屋面以下、马道上方设置喷头保护马道区域。

11. 消防水炮系统

（1）系统设置。航站楼超过12m以上高大空间采用水炮系统，包括自动扫描射水高空水炮系统（小炮系统）和固定消防炮灭火系统（大炮系统）。其中主航站楼设置大炮系统，每门水炮设计流量20L/s，最大射程50m，系统设计最多可同时开启2门水炮灭火。指廊及连廊采用小炮系统，每门水炮设计流量5L/s，最大射程20m~25m，系统设计最多可同时开启6门水炮灭火。

（2）消防泵组。本系统消防泵组及管网均独立设置，消防泵组设于交通中心地下层的消防泵房内。采用三台水炮加压泵，两用一备。稳压泵设于主楼标高24.175层高位水箱间内。系统主管环状布置。各指廊主管接入处均设可调式减压阀。

（3）系统控制。水炮控制系统现场画面可视，报警图象多重备份，报警同时进行语音通讯。一旦发生火灾，信息处理主机发出报警信号，显示报警区域的图像，并自动开启录像机进行记录，同时通过联动控制台采用人机协同的方式启动自动消防炮进行定点灭火。系统有三种控制方式：自动控制、消防中心手动控制、现场应急手动控制。

12. 高压细水雾灭火系统

（1）系统设置。地下管廊电舱部分、发电机房、TOC操作大厅、GTIC分控中心等部位设置高压细水雾系统。采用开式分区应用系统，各分区由区域控制阀控制。

（2）消防泵组。在主楼首层及东西指廊各设置一套高压细水雾泵组，三套泵组可互为备用。

（3）系统控制。在准工作状态下，从泵组出口至区域阀前的管网由稳压泵组维持压力1.0~1.2MPa，阀后空管。发生火灾后，由火灾报警系统联动开启对应的区域控制阀和主泵，喷放细水雾灭火；或者手动开启对应的区域控制阀，管网降压自动启动主泵，喷放细水雾灭火。经人员确认火灾扑灭后，手动关闭主泵和区域控制阀，火灾报警系统复位，管网恢复，系统复位。系统具备三种控制方式：自动控制、手动控制和应急操作。

13. 气体灭火系统

（1）系统设置。重要设备用房、强弱电机房等不宜水消防的部位，采用七氟丙烷气体灭火系统。防护区较集中区域采用组合分配式管网灭火系统，共设置42个组合分配系统，每个系统防护区不超过8个。分散的防护区采用预制灭火系统。

（2）系统构成。组合分配系统由高度智能型火灾自动报警探测、控制系统、气体灭火系统设备及灭火剂输送管道组成。预制灭火系统由火灾自动报警系统、灭火控制系统和灭火无管网系统组成。

（3）系统控制。组合分配系统具有自动启动、手动启动和机械应急操作三种启动方式。

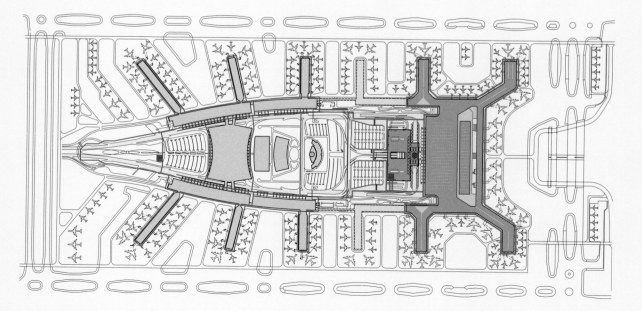

航站区消防总图
Fire Control Master Plan of Terminal Area

— 一号航站楼消防管网
　　Fire Control Pipeline Network of T1

— 二号航站楼消防管网
　　Fire Control Pipeline Network of T2

■ 消防水池
　　Fire Water Pool

8. Washroom Plumbing

Washroom water is supplied by zone to ensure the safety of the drainage system. The branched water supply system is combined with annular one for more stable and comfortable water supply. Maintenance access is provided on the back wall of each row of washroom cubicles for easy construction and maintenance.

9. Fire Hydrant System

(1) Outdoor fire hydrant system. The outdoor firefighting system includes airside and landside parts. The outdoor firefighting system on the air side and the movement area firefighting system share one same pipe network. For the landside outdoor firefighting system, two DN300 pipes are introduced from the DN500 water supply pipe network of the terminal area to form an annular outdoor firefighting pipe network.

(2) Indoor fire hydrant system. The indoor and outdoor firefighting systems of T2 are provided separately. The indoor firefighting systems (including indoor fire hydrant system, automatic sprinkler system, water curtain separation and protection system) share one same pressure pipe network.

10. Automatic Fire Sprinkler System

(1) Automatic fire sprinkling system is provided except for rooms and positions such as lofty spaces, stairwells and less than 5m² washrooms that are not prone to fire or not applicable for fire extinguishing by water. Fire design in ordinary area is based on medium-risk Grade I, and in the parking building of the GTC medium-risk Grade II.

(2) The pressure pump set and pressure pipe network is shared by the indoor fire hydrant system. Several alarm valve rooms are provided by zone in the terminal. In each alarm valve room, two water supply pipelines are introduced from the fire-fighting pressure pipe network to the alarm valve bank. Maintenance valve, electric gate valve and check valve are installed at each introduced pipeline. Water flow indicators are installed on each floor and in each fire zone. The water flow indicator signals are displayed in the fire control center. Fire pump adapters are provided by zone.

(3) Selection and layout of water nozzles. Quick response nozzles are used. Decorative downward sprinkling nozzles are hidden in suspending ceiling. Upward nozzles are installed in ceiling higher than 800. The action temperature of sprinklers is 93℃ in kitchen, 79℃ in ceiling, and 68℃ in other locations. Sprinklers are installed below the metal roof and above the catwalk to protect the catwalk area.

11. Fire Monitor System

(1) System setting. Fire monitor systems are used in tall spaces over 12m in the terminal, including automatic scanning and jetting fire monitor system (small monitor system) and fixed fire monitor system (big monitor system). In the main building, big monitor system is provided, with the designed flow of each monitor being 20L/s and the maximum jetting distance being 50m. The system is planned such that maximum 2 monitors may operate simultaneously to extinguish the fire. Small monitor systems are used for the piers and the connecting corridors with the designed flow of each monitor being 5L/s and the maximum jetting distance being 20-25 m. The system is planned such that maximum 6 monitors may operate simultaneously to extinguish the fire.

(2) Fire pump set. The fire pump set and pipe network of this system are provided independently. The fire pump sets are located in the fire pump room in the basement of the GTC. Three fire monitor pressure pumps are used, with two for service and one for standby. The pressure stabilizing pump is placed in the high-level water tank room at Level 24.175m in the main building. The system main pipe is in annular layout. Adjustable pressure relief valves are provided at the main pipe introducing positions of all piers.

(3) System control. The scene images of the fire monitor control system are made visible, and the alarm image can be generated into multiple copies for backup purpose. Voice communication works as well during the alarming process. In case of fire, the information processing host computer sends out an alarm signal, displays the image of the alarmed area, and automatically turns on the video recorder to record the fire field. At the same time, the automatic fire monitor is activated from the linkage console via man-machine collaboration for fixed-point firefighting. The system allows three control modes: automatic control, manual control from the fire control center and manual emergency control at the field.

12. High Pressure Water Mist Fire-extinguishing System

(1) System setting. High-pressure water mist systems are provided in the electrical cabins of utility tunnel, generator room, TOC operation hall, GTIC sub-control center, etc. It adopts open zone application system that allows zoned control by zone control valves.

(2) Fire pump set. One set of high-pressure water mist pump set is respectively provided on F1 of the main building and the east and west piers. The three sets of pump sets mutually back up each other.

(3) System control. In the quasi-working state, the pipe network between the outlet of the pump set and the front of the zone valve is maintained by pressure stabilizing pump at a pressure of 1.0-1.2 MPa, while empty pipe behind the valve. In case of a fire, the corresponding zone control valve and main pump are started in a coordinated manner by the fire alarm system to spray fire-extinguishing water mist; or the corresponding zone control valve is started manually so that the pipe network depressurizes for auto start of the main pump to spray water mist for fire extinguishing. After fire suppression is confirmed, the main pump and the zone control valve are manually closed, the fire alarm system is reset, the pipe network is restored and the system is reset. The system allows three control modes: automatic control, manual control and emergency operation.

13. Gas Fire Extinguishing System

(1) System setting. HFC-227ea fire-extinguishing system is adopted for important MEP rooms, HV and ELV rooms and other positions where fire-fighting with water is unacceptable. 42 combined distribution type pipe network fire-extinguishing systems are provided in concentrated protection zones, with each system covering maximum 8 protection zones. Prefabricated fire extinguishing system is used in scattered protection zones;

(2) System composition. The combined distribution system consists of highly intelligent automatic fire alarm detection and control system, gas fire-extinguishing system equipment and fire extinguishing agent delivery pipeline. The prefabricated fire-extinguishing system consists of automatic fire alarm system, fire extinguishing control system and pipeline-free fire extinguishing system.

(3) System control. The combined distribution system allows three start modes: automatic start, manual startg and mechanical emergency operation.

7.3.3 建筑电气设计

7.3.3.1 负荷计算

1. 负荷等级。本工程为Ⅲ类及以上民用机场航站楼,详见"负荷分类表";
2. 负荷容量。①本工程二号航站楼总计算负荷为56499kW（不含设于交通中心的制冷机房用电），变压器总装机容量为111700kVA，平均负荷率为48.5%。按远期规划规模，包含空侧设备与行李系统安装容量，变压器安装指标为161.6VA/m²。②设备中心10kV空调制冷主机设备安装容量为20096kW，按功率因数0.92折算变压器装机容量为21843kVA；空调冷源其他设备（含380V空调制冷主机与冷冻泵、冷却泵、冷却塔及其配套设备）计算负荷为8990kW，变压器装机容量为14400kVA，平均负荷率为67.9%。设备中心空调制冷设备折算变压器总装机容量为36243kVA。③航站楼整体变压器安装指标为203.6VA/m²。④项目（含交通中心）消防设备计算负荷为7693kW，市电停电时需要发电机确保的一级负荷计算负荷为8409kW；其中航站楼消防设备计算负荷为6352kW，市电停电时需要发电机确保的一级负荷计算负荷为7388kW。

7.3.3.2 变电所设置

二号航站楼共设置18个10kV/0.4kV变电所，其中16个公用变电所、2个行李系统专用变电所；设备中心设置6个10kV/0.4kV变电所（其中4个制冷机房专用变电所、2个停车楼公用变电所）。

7.3.3.3 10kV供电系统

1. 航站楼内10kV供电系统。二号航站楼内设置6个区域主变电所，共引入12路10kV电源供电。每个区域主变电所分别从新建110kV变电站不同变压器母线段引来两路10kV电源，两路10kV电源同时工作，100%互为备用。原则上各变电所变压器两台一组，采用单母线分段运行方式，每组变压器两路10kV电源分别直接引自同一个区域主变电所不同的10kV母线段。
2. 设备中心10kV供电系统。设备中心设置2个区域主变电所，共引入6路10kV电源供电。每个区域主变电所分别从新建110kV变电站不同变压器（变电站共设3台主变压器）母线段引来三路10kV电源，三路电源两用一备，备用回路能备用其中一路主用的100%负荷。每台10kV高压空调制冷机组分别由区域主变电所直接引一路10KV电源至机组自带的控制柜进线端，高压空调制冷机组采用变频控制的方式，启动电流不大于正常运行电流的5倍。控制柜分主楼系统与指廊系统分别集中安装于控制室内。各变电所变压器两台一组，采用单母线分段运行方式，每组变压器两路10kv电源分别直接引自同一个区域主变电所不同的10KV母线段。
3. 备用电源10kv供电系统。在西指廊设备中心设置5台常用功率为1800kVA的高压发电机组作为备用电源，并在东、西连接楼分别设置一个10kv备用电源二级配电室，向各变电所备用变压器供电。用于保障二号航站楼及交通中心内特别重要负荷、消防设备、弱系统、应急照明等负荷在市电故障时的连续供电。备用电源系统由一路市电与一路发电机电源供电，其中市电从新建机场北110kV变电站直接引来。平时由市电供电，当其市电故障时或发生火灾时，向发电机主控柜发送启动发电机信号。发电机备用电源系统送电后，市电电源进线柜开关分闸，发电机出线柜开关自动合闸，发电机备用电源投入使用。当市电电源回路恢复供电时，手动断开发电机出线柜开关，手动合上市电进线的电源开关，转换成正常情况下的供电方式，采用手动方式将发电机逐台退出运行。在东、西连接楼的10kv备用电源二级配电室采用两路电源一用一备的方式供电，其中主供回路由西设备中心备用电由交通中心备用电源系统供电，当该电源故障时，经延时后主供电源进线柜开关自动分闸，备供电源进线柜开关自动合闸，由机场南110kV变电站电源供电。当主供电源回路恢复供电时，视情况手动断开备用电源进线柜开关，手动合上主供电源进线的电源开关，转换成正常情况下的供电方式。

7.3.3.4 0.4kV配电系统

1. 系统运行方式。①常规变压器：除E5A及W5A变电所的B3变压器外，每两台变压器一组，采用单母线分段运行方式，正常时两台变压器各带约50%负荷，平均负荷率小于50%。当一台变压器故障时，由另一台变压器带两段母线上全部负荷。当两路市电进线中的一路或一台变压器故障时，低压母联开关采用自投不自复方式，自投时间滞后于进线断路器失压脱扣时间不小于0.5s。两路进线与联络断路器之间采取电气联锁措施，保证三台断路器不能同时处于合闸状态。E5A及W5A两变电所B3变压器平时独立运行，当该变压器故障退出运行时，手动分闸该进线开关后，合上与B2间设联络开关。②备用变压器：采用单母线运行方式，平时变压器负荷率小于50%。平时正常运行时由备用电源系统10kV市电电源供电，当市电故障或火灾时由发电机10kV备用备用电源供电；平时当备用变压器故障或检修时，该0.4kV母线段由变电所内市电母线段供电。
2. 各级负荷供电方式。①一级负荷中特别重要负荷：民航类弱电采用两路电源、两回线路（一路市电、一路备用电源供电系统）引至UPS室内并设置进线开关，并保证任一回路在故障情况下，另一回路能承担100%负载容量（UPS配电采用的是双总线结构）；其他弱电系统采用两路电源、两回线路（一路市电、一路备用电源供电系统）末端自动切换，并设置在线式UPS供电。②一级负荷中的消防负荷：采用两路电源、两回线路（一路市电、一路备用电源供电系统）末端自动切换。③一级负荷中的非消防负荷：大空间照明采用两路电源、两回线路（两路市电）各带一半负荷的供电方式；其他一级负荷采用两路电源、两回线路（两路市电）末端自动切换。④二级负荷：采用两路电源、一回线路供电。
3. 配电方式。本设计采用放射与树干相结合的配电方式，分区设置配电竖井与楼层配电间。根据负荷类别及管理要求，分类设置公共区照明、公共区地面用电、应急照明、办公区用电、公共区域送排风动力、空调通风动力、生活水动力、排水动力、自动扶梯与步道、电梯、消防动力、弱电系统UPS、航显标识用电、联检、贵宾、两舱休息室、广告、商业、机坪各种用电等专用低压回路与区域配电箱（柜）。其中设备机房及容量大的设备采用放射式供电，其他设备以配电数据为单位采用树干上供电。应急照明以竖井为单位在区域配电间内集中设置应急电源装置（EPS）供电，EPS持续供电时间为60min，应急供电转换时间不大于0.5s。

变电所点位
Substation Point Positions

● 区域变电所兼分变电所8个
 8 Regional & Transformer Substations

● 分变电所14个
 14 Transformer Substations

7.3.3 Building Electrical System

7.3.3.1 Load Calculation

1. Load level. This project is a civil airport terminal of and above Class III. Refer to Load Classification Table for details.
2. Load capacity. ① T2 has a total calculation load of 56,499kW (excluding the electrical demand of the refrigerator room in GTC), a total installed transformer capacity of 111,700kVA and the average load rate of 48.5%. The long-term planned capacity includes the installed capacity of airside equipment and baggage system, with the transformer installation index of 161.6VA/m². ② The installed capacity of 10kV AC refrigerators in the MEP Center is 20,096kW and the installed capacity of transformers converted as per power factor of 0.92 is 21,843kVA. The calculation load of other equipment for AC cold source (including 380V AC refrigerator and refrigeration pump, cooling pump, cooling tower and their supporting equipment) is 8,990kW, the installed capacity of transformers is 14,400kVA, and the average load rate is 67.9%. The converted total installed capacity of transformers for AC refrigeration equipment in the MEP Center is 36,243kVA. ③ The overall transformer installation index of the terminal is 203.6VA/m². ④ The calculated load of firefighting equipment (including GTC) for the project is 7,693kW, and the calculated primary loads to be supplied by power generators, in case of commercial power supply failure, is 8,409kW; among them, the calculated load of firefighting equipment in the terminal is 6,352kW, and the calculated primary loads to be supplied by power generators, in case of commercial power supply failure, is 7,388kW.

7.3.3.2 Substations

A total of eighteen 10kV/0.4kV substations are provided in T2, including 16 public substations and 2 dedicated for baggage system. Six 10kV/0.4kV substations (including 4 dedicated for refrigerator rooms and 2 public ones for the parking building) are provided for the MEP Center.

7.3.3.3 10kV Power Supply System

1. 10kV power supply system in T2. Six zoned main substations are provided for T2, where twelve 10kV power supply circuits are led in. Two 10kV power supply circuits are led into the main substation of each zone respectively from different transformer bus sections of the newly-built 110kV substation. Such two 10kV power supply circuits operate simultaneously and are 100% mutually backed up. In principle, two transformers in each substation are in a group and operate in the mode of single bus section, and two 10kV power supply circuits of each transformer group are directly introduced from different 10kV bus sections of the main substation respectively in the same zone.
2. 10kV power supply system of the MEP Center. The MEP Center is equipped with 2 zoned main substations, where six 10kV power supply circuits are led in. To each zoned main substation, three 10kV power supply circuits are introduced from the bus sections of different transformers in the newly-built 110kV substation (three main transformers are installed in such substation). Two of the three power supply circuits are in service and one for backup. The backup circuit can back up 100% load of either of the two main service circuits. For each 10kv high-voltage AC refrigeration unit, one 10kV power supply circuit is introduced from the zoned main substation to the incoming line terminal of the control cabinet of the unit. The high-voltage AC refrigeration unit is equipped with frequency conversion control and the starting current is not more than 5 times of the normal operating current. The control cabinets serve the main building and piers separately and are centralized in the control room. Two transformers in a group are provided in each substation and operate in the mode of single bus section, with the two 10kV power supply circuits of each transformer group directly introduced from different 10kV bus sections of the main substation respectively in the same zone.
3. Backup 10kV power supply system. Five high-voltage generator sets with normal power of 1,800kVA are provided in the MEP Center of the west pier as backup power source, and a secondary distribution room of 10kv backup

负荷等级 Load Level	负荷类别 Load Category
一级负荷中特别重要负荷 Particularly Important Primary Load	应急照明；边检、海关的安全检查设备；航班信息、显示及时钟系统；航站楼、外航驻机场办事处中不允许中断供电的重要场所用电负荷；重要电子信息机房、防灾中心及分控室（包括消防中心、分控室）、集中监控管理中心、应急指挥中心；安防系统、信息及弱电等系统专用电源等。 Emergency lighting; security check equipment of frontier inspection and customs; flight information, display and clock system; important places in the terminal and foreign airline offices where power supply interruption is not allowed; important electronic information machine room, disaster prevention center & sub-control room (incl. fire control center, sub-control room), centralized monitoring & management center, emergency command center; dedicated power supply for security, information, ELV and other systems.
一级负荷 Primary Load	消防水泵、消防风机、消防电梯及其他消防设备设施用电；生活水泵、潜污泵、雨水泵；行李系统；出发大厅照明，到达大厅照明，联检大厅照明，值机及候机厅照明，其他公共区域照明，公共区域送排风系统设备，普通客梯；海关、边检、检验检疫等区域用电，贵宾用电等。 Fire pump, fire blower, fire elevator and other fire equipment & facilities; domestic water pump, submersible sewage pump, rainwater pump; baggage system; lighting in departure hall, arrival hall, CIQ hall, check-in and waiting hall as well as other public areas; air supply and exhaust system and ordinary passenger elevators in public areas; customs, frontier inspection, CIQ and other areas; VIPs.
二级负荷 Secondary Load	公共场所空调系统设备、自动扶梯、自动人行道；货梯，贵宾厨房等动力及商业用电；广告用电，上述一级负荷外的其他用电。 AC system equipment, escalators, and APMs in public spaces; freight elevators, VIP kitchen and electricity for other commercial purposes; advertising, other loads excluded in the aforesaid category of primary load.

负荷分类表
Load Classification Table

power supply is respectively provided in the east and west connecting buildings to supply power to the backup transformers of each substation. The backup power supply system is used to guarantee continuous power supply to special important loads and loads of firefighting equipment, ELV system, and emergency lighting in T2 and GTC in case of commercial power failure. The backup power supply system is powered by one circuit of commercial power source and one circuit of generator power source, with the commercial power supply directly introduced from the new 110kv substation in the north of the airport. At ordinary time, the power is supplied by the commercial power source, and in case of commercial power failure or a fire, the signal for starting the generators is sent to the main control cabinet of the generators. When the backup generator power supply system starts supply of power, the switch in the commercial power incoming cabinet is turned off and the switch in the generator outgoing cabinet is automatically turned on to put the backup generator power supply into operation. When the commercial power supply is resumed, the switch of the generator outgoing cabinet is manually disconnected and the power switch on the commercial power incoming line is manually turned on to resume normal power supply, and the generators are manually shut down one by one. In the secondary distribution room of 10kv backup power supply in the east and west connecting buildings, two power supply circuits with one for service and one for backup are adopted. The main service circuit is supplied by the backup power source of the west MEP Center, and the backup service circuit by the backup power supply system of the GTC. In case of power source failure, the switch of the main power supply incoming cabinet is turned off automatically after a delay and the switch of the backup power supply cabinet is automatically turned on for power supply from the 110kV substation in the south of the airport. When the main power supply circuit resumes power supply, the switch of the backup power supply incoming cabinet is manually disconnected and the power switch of the main power supply incoming line is manually turned on as the case may be to resume the normal power supply mode.

7.3.3.4 0.4kV Power Distribution System

1. System operation mode. ①Conventional transformers: except B3 transformers in E5A and W5A substations, every two transformers are in a group and operate in the mode of single bus section. Normally, each of the two transformers supplies to about 50% load with the average load rate less than 50%. When one of the transformers fails, the other transformer bears all the loads on the two bus sections. When one of the two commercial power supply incoming lines or one of the transformers fails, the low-voltage bus coupler switch automatically switches in, but not automatically resets. The automatic switch-in lag time is not less than 0.5s after the voltage loss tripping time of the incoming circuit breaker. Electrical interlocking measures are provided between the two incoming lines and the interconnecting circuit breakers to prevent that the three circuit breakers are on at the same time. B3 transformer in E5A and W5A substations operates independently at ordinary time. When the transformer fails and exits and after manually switching off the incoming switch, the interconnecting switch with B2 is manually switched on. ②Backup transformer operates in the mode of single bus with the normal load rate less than 50%. During normal operation, power is supplied from the 10kV commercial power source of the backup power system; and when the commercial power supply fails or a fire occurs, the power is supplied from the backup 10kV power source of generators; at ordinary time when the backup transformer fails or is in maintenance, the 0.4kV bus section is powered by the commercial bus section in the substation.

2. Mode of power supply to each level of loads. ①Particularly important primary loads: For civil aviation ELV system, two lines of power sources and two-circuit lines (one line of commercial power source and one line of backup power supply system) are used for civil aviation system and connected into UPS room, where incoming switches are installed, to ensure that one circuit can bear 100% load (UPS power distribution in dual-bus structure) when the other circuit fails. For other ELV systems, two lines of power sources and two-circuit lines (one line of commercial power source and one line of backup power supply system) are used with for automatic terminal switching and on-line UPS is provided for power supply. ② Fire-fighting primary loads: two lines of power sources and two-circuit lines (one line of commercial power source and one line of backup power supply system) with automatic terminal switching are adopted. ③ Non-firefighting primary loads: two lines of power sources and two-circuit lines (two lines of commercial power supply) are used for large space lighting with each line bearing half of the loads; for other primary loads, two lines of power sources and two-circuit lines (two lines of commercial power supply) with automatic terminal switching are used. ④Secondary loads: two lines of power sources and one-circuit lines are used for power supply.

3. Power distribution method. Radial and trunk-type power distribution are provided, with power distribution shafts and floor power distribution rooms provided by zone. According to the load type and management requirements, special low-voltage circuits and zone distribution boxes (cabinets) are provided for public area lighting, public area ground power consumption, emergency lighting, office area power consumption, air supply and exhaust, AC and ventilation, domestic water, water drainage, escalators and walkways, elevators and firefighting in the public area, UPS of ELV system, flight information display and signage, CIQ, VIP area, first class/business class lounge, advertising, retails, apron, etc. Radial power supply is adopted for MEP rooms and equipment with large capacity, while trunk-type power supply is adopted for other equipment based on the unit of power distribution data. For emergency lighting, emergency power supply (EPS) devices are centralized in the zone power distribution room in the unit of vertical shaft. The continuous power supply time of EPS is 60min, and the emergency power supply switching time is no more than 0.5s.

7.3.4 通风空调设计

7.3.4.1 概述
1. 空调系统

二号航站楼及交通中心停车楼主要公共区域采用集中式水冷中央空调系统作为夏季空调，交通中心的变配电房采用直接蒸发小型风冷螺杆机提供冷源，一至三层分散零星小办公室和TIC分控中心（监控大厅）采用分体空调，部分较远的办公室采用风冷智能多联空调系统。除特殊功能用房外（如登机桥、贵宾、两舱、国际计时旅馆等），其他区域冬季不设空调采暖。特殊功能用房采用智能变频多联冷暖空调系统，指廊部分两舱采用风冷热泵机组专门负责冬季供暖空调。二号航站楼及交通中心共设置冷冻水系统4个，制冷机房2个。分别设在交通中心外东、西两侧地下负二层。总冷负荷约为119553kw（34000冷吨），另加夜间运行小主机冷量，总装机冷量为35680冷吨。建筑单位冷指标为175W/m²。

节能措施：办票岛大厅、旅客出发厅、到达厅、行李提取大厅等大空间采用全空气系统。在供冷期根据室内外的焓值确定新风量，在过渡季节，尽量采用自然通风，根据室内外压差开启电动排烟窗，把热气排出室外。

排风热回收技术：办公室、贵宾区、两舱和计时旅馆等功能房间采用热泵式热回收型溶液调湿新风机组，当系统新风量≥3000cmH，并且热回收的排风量≥70%时提取排风的冷量，以达到节能的目的。

空调部分负荷运行策略：首先制冷系统采用了大小机组相结合配置，调节性能好，能有效地适应负荷变化的要求，防止产生浪费现象。其次项目采用冷源群控系统，根据冷量控制冷水机组的启停数量及其对应的水泵、冷却塔的启停数。此外项目空调冷冻水循环系统采用变流量控制，并且对二次循环水泵采用变频调速和台数控制。

2. 其他系统

（1）通风系统。车库、设备用房、卫生间、行李库、吸烟室、综合管沟等区域设平时通风系统。
（2）防排烟系统。大楼防排烟系统按照国家现行消防规范和性能化要求进行设计。
（3）人防战时通风系统。交通中心及停车楼负一、负二层设置了人防掩蔽场所，相应设计人防战时通风系统。

3. 空调末端

本工程最大空调器风量为13万m³/h，风量大于或等于3万m³/h的空调器采用变频空调器，风量大于或等于5万m³/h的空调器采用组合式空调器。新风空调器采用初效过滤器+静电除尘空气净化装置；带回风的空调器采用初效G4+F7中效过滤。

本工程风机最大风量15.5万m³/h。行李库采用上、下排风相结合的方式，消防时关闭下排风。小弱电机房采用风机盘管加排风系统。

7.3.4.2 设计特点

1. 本工程面积较大，工程总共分成四个制冷系统，使系统的水力平衡更易实现，系统的使用安全可靠，控制及调试快速便捷。
2. 制冷系统采用大小机组相结合的配置，调节性能好，能有效地适应负荷变化的要求。
3. 冷却塔采用混凝土水池，池与池之间采用4根直径为1000mm的连通管，连通管处的水池深1.5m，其余为1m深，较好平衡了冷却塔之间的水系统。
4. 为了平衡水路，各末端空调器及风机盘管组群支路均采用了静态平衡阀+动态压差平衡阀，既考虑了阻力平衡又考虑了系统运行的安全性和可靠性。
5. 全年需要空调确保的房间如贵宾室、两舱和计时旅馆等则采用风冷智能多联冷暖式空调方式，夏季制冷，冬季供暖。系统控制灵活，运行费用低。
6. 卫生间设下排风口和上排风口结合的方式，排风效果相比仅设上排风口更佳。
7. 机场制冷机房的空调水管通过综合管廊进入航站楼，然后再接空调水立管至各个区域。既方便人员检修，又保证了航站楼区域室内净高要求。
8. 交通中心地下车库进风由自然进风井、天井和车道自然补进，地面的车库进风由周边与室外相通的栏杆及天井自然补进；通风排烟系统相结合，排风采用竖井排风方式，水平方向没有风管，提高了使用空间净高，并节约了投资。
9. 结合气流组织的合理性和建筑的美观要求，处理后的空调风通过侧送风口、下送旋流风口或风亭等送风方式送入室内。
10. 变配电房采用两套降温系统：当室外温度≤31℃且室内温度≤35℃时，采用通风降温；当室外温度>31℃或室内温度>35℃时，采用空调降温。节省了运行费用。
11. 考虑到机场部分区域（如部分办公和电气设备用房）需要24h使用，但冷量相对较小，空调水管系统专门设计了一套小管路系统，每个制冷系统设计一台小主机专供夜间运行。为机场夜间安全、可靠运行做了细致考虑。

7.3.5 建筑智能化设计

7.3.5.1 设计内容

建筑智能化系统主要包括建筑设备管理系统、能效管理系统、机电设备信息管理平台几大部分，另外还有配套为这几个系统服务的设备管理网布线系统、计算机网络系统及配电、防雷接地工程等。不包含民航类弱电、信息、安防系统。

7.3.5.2 系统组成
1. 各系统功能及组成

建筑设备管理系统主要对空调、给排水、照明、电梯、扶梯、步道等机电设备进行监控和管理，并可进一步通过集成实现各系统与航班信息的联动控制。包括的子系统有：建筑设备监控系统、冷源群控系统、智能照明控制系统、专用设备监控系统、建筑设备集成管理系统。

能效管理系统主要对电力系统实施自动监测，对用户的水、电能耗进行统计，在此基础上建立全景数据库和能源消耗评价体系。包括的子系统有：电力自动监控系统、自动计量系统、EPS电源监测系统、能效管理平台。

本项目设置一个机电设备信息管理平台，对各主要机电设备设置统一的编码并以二维码为展示载体，实现各机电系统信息共享与资源整合，打通建筑设备管理与能效管理系统，便于机电系统相关信息的跟踪、关联，提高设备的安全性和可维护性。

为满足以上两大系统的网络通信需要，设置一套专用的建筑设备管理计算机网络及布线系统。

2. 监控机房的设置

建筑智能化系统总控中心设于二号航站楼主楼西南区二层的设备管理中心机房，各系统主服务器均设于该机房内。根据建筑分区和管理需要，分别设置19个设备管理机房，其中部分机房兼具分控室功能。

7.3.5.3 建筑设备管理系统
1. 建筑设备监控

本系统对普通空调通风、给排水及污水、溶液式新风空调器、太阳能热水、雨水回收利用、电动遮阳百叶、气动排烟窗、自动门等进行自动监测或控制。

系统由服务器、工作站、网络控制器、现场控制器、各类传感器及执行机构、控制层/管理层网络以及操作系统软件和应用软件等构成，并与建筑设备管理系统进行集成。主服务器设于设备管理中心机房内，服务器采用双机冗余技术。在设备管理中心机房、兼具分控功能的设备管理机房、TOC监控中心设置管理工作站；系统网络控制器（NC）、现场控制器（DDC）设于各设备管理机房或末端设备间内。

系统监控总点数为11555点，通信网络分为两级结构：管理层及控制层。管理层网络即为设备管理网，采用千兆主干以太网，采用客户机/服务器数据处理模式，支持TCP/

IP通信协议；控制层网络采用BACnet总线连接各个网络控制器、现场控制器，控制器之间可通过控制层网络实现点对点通信。

2. 智能照明控制系统

智能照明控制系统对旅客办票大厅、安检大厅、候机厅、到达厅、行李提取厅、贵宾室、卫生间等公共区域及办公区走道等区域的照明进行控制及管理。系统由管理服务器、工作站、网关、时钟管理器、电源模块、可编程开关控制器、可编程调光控制器、可编程控制面板、液晶触摸屏、输入模块、照度探测器、人体感应器及管理软件等部件组成。主服务器设于二号航站楼主楼二层设备管理中心机房内，服务器采用双机热备冗余技术。

3. 建筑设备集成管理系统

采用建筑设备集成管理系统将二号航站楼的建筑设备监控系统、冷源群控系统、智能照明控制系统、各专用设备监控系统、能效管理系统等以交换式以太网组网进行中央集成，生成建筑运行管理所需要的综合数据库，从而对所有全局事件进行集中管理，并实现各子系统之间的信息共享和集中的设备监控、报警管理和联动控制功能。

7.3.5.4 能效管理系统

1. 电力监控系统

（1）电力自动监控系统对二号航站楼的高低压配电系统、变压器、直流屏等实施自动监测（高压系统含保护及控制），同时对变配电房的门开关状态进行监测。系统数据上传至能效管理软件平台进行分析处理。系统采用间隔层、站级层和网络层三层网络结构；

（2）实现功能：①对10kV高压配电系统实行自动监视、继电保护和测量。②对0.4kV低压配电系统、变压器、直流屏等电力设备实行自动监视和测量。③对电力系统的运行参数进行自动采集、分析及实时监测，及时消除故障隐患，并进行集中管理。④对变配电房门开关状态进行监视。⑤各种仪表需要有开关分合闸状态、故障状态信号输入端子。⑥联络开关处智能仪表应该具有双向电度测量功能，系统读取并监视。⑦变压器进线开关处智能仪表应设置电流超限"预报警"功能，电流超限"报警"功能。

2. 自动计量系统

自动计量系统通过对二号航站楼内的出租商铺、航空公司用房、驻场单位用房、公用卫生间、部分楼层总表等用户的用水量、用电量数据进行远程采集。系统数据同时上传至能效管理软件平台进行分析处理。系统由管理服务器、工作站、通信管理器、总线通信数字电度表、总线通信数字水表等组成。

自动计量系统数据通过系统管理软件作初步的统计分

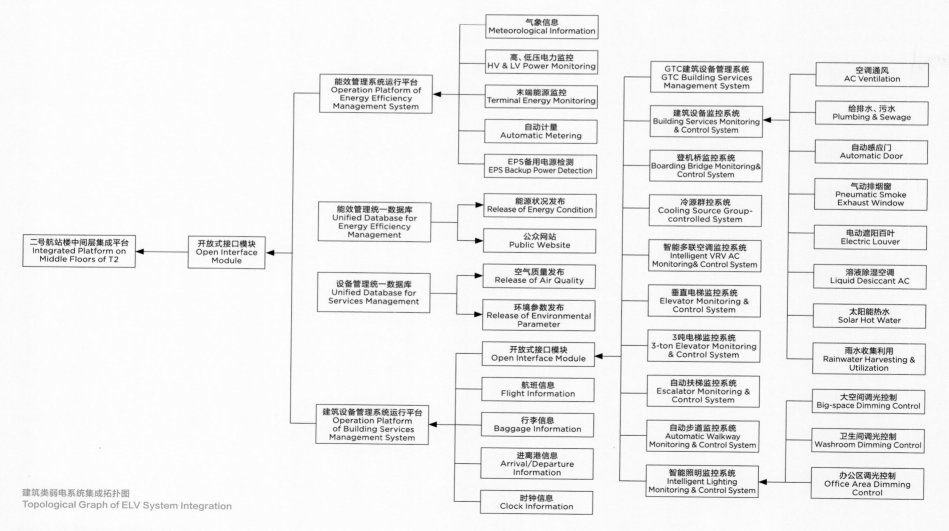

建筑类弱电系统集成拓扑图
Topological Graph of ELV System Integration

析，形成报表用于日常的管理和收费。系统数据同时上传至能效管理系统，由能效管理系统平台结合电力监控系统数据一并进行统一分析，最终形成整个航站楼的能耗数据。

3. EPS电源监测系统

EPS电源监测系统通过对航站楼配电间的EPS设备通过标准通信接口采集设备运行数据，上传至能效管理系统平台，实现对设备的远程监控管理。系统采用二级结构。上层为通信管理器，设于各区设备管理机房，与自动计量系统通信管理器共用，向上接入设备管理网与能效管理系统服务器进行通信，向下以MODBUS总线方式与各配电间的EPS直接通信。本系统不设置独立的后台软件，各EPS运行数据直接上传至能效管理系统，由能效管理系统平台软件进行统一监测和分析。

4. 能效管理平台

系统在电力监控、自动计量系统软件的基础上，进一步对二号航站楼内的水、电等能耗进行采集、分析和对管理模块进行集成与整合，采用通用数据模型建立全景数据库，并通过专家系统对能耗系统和环境数据实行实时监控、统计分析、预测分析等。

7.3.5.5 机电设备信息管理平台

1. 为了加强二号航站楼内机电设备（系统）的信息管理，通过一体化平台对设计、生产、安装、调试、验收及使用维护全生命周期信息的跟踪监管，本项目对机电用房与机电设备采用统一的编码技术，设置一套机电设备信息管理平台系统。

2. 机电设备信息管理平台的建设内容：
（1）建立二号航站楼专有的命名与编码机制。对机电设备的分类管理，统一命名与编码机制，确保命名与编码规则唯一化；同时，对具体的房间号和位置号进行编码；
（2）二维码管理技术的全面普及在机电设备上挂牌管理，扩充设备可查看信息。
（3）设备生命周期全面掌控。实现设备从设计、生产、安装及使用的全过程的有效跟踪及信息一致性，便于相关信息的跟踪，提高设备的安全性和可维护性。
（4）先进、高效、易扩展 的统一信息平台：建立统一的管理平台及数据库，并与航站楼内各机电管理系统（建筑设备管理系统、能耗管理系统）平台对接，实现数据共享。

7.3.5.6 智能化设计特点

1. 分散控制、集中管理。在二号航站楼内根据建筑分区和管理需要，分别设置16个设备管理机房，其中部分机房兼具分控室功能；服务于航站楼大跨度总控与分控的管理机制。

2. 专线专网、高速通信。为满足各建筑弱电系统（火灾自动报警系统除外）的网络通信需要，设置一套专用的建筑设备管理计算机网络及布线系统。并以单模光纤作为布线主干传输介质，以双万兆交换机作为核心网络，解决了航站楼大跨度通信的难点。

3. 全面监管、高度集成。将二号航站楼的各类空调、各类给排水、照明、电梯、扶梯、步道、气动排烟窗、自动感应门、电动遮阳百叶等机电设备在建筑设备管理平台上全面监管，并通过进一步集成实现各系统的联动控制，提升航站楼的管理水平。

4. 优化能效、智能管理。能效管理系统对电力系统、末端水、电能耗进行统计，在此基础上建立全景数据库和能源消耗评价体系，对建筑的整个能耗进行分析，提出节能降耗的技术和管理措施，达到节能降耗目的。同时，为便于管理航站楼内的机电设备，设置机电设备信息管理平台，为各主要机电设备设置统一的编码并可以二维码为展示载体，实现建筑设备管理及能效管理系统所监管设备具有唯一性，也可方便两个系统数据的互联。

5. 因地制宜、多彩调光。根据二号航站楼出发/到达大厅、值机岛、贵宾厅、后勤办公区等多种场所的特点，选用了0~10V调光、DMX512、DALI调光、开关控制等多种技术；特别在值机岛上空配合WRGB投光灯，采用DMX512舞台灯光控制方式，实现色彩变幻的调光效果，增加旅客的现场体验感。

7.3.6 设备管廊设计

7.3.6.1 设计特点

1. 路由规划特点。根据建筑布局及特点、结合各专业设备机房所在位置进行规划，既保证各类管线的安全和便于安装维护，也做到尽可能利用设备管廊的面积。
2. 水电分仓理念。设备管廊分水仓和电仓，避免水对电气管线的影响，解决消防问题，同时也避免水电管线交织带来施工及管理的麻烦。
3. 断面设计。结合水、电管线敷设，结合施工空间、检修空间、防火疏散综合考虑，最终确定断面的尺寸。

7.3.6.2 设计难点

设计难点主要为电仓与水空仓管线交叉处理。电仓内管线较多、桥架层数较多，在交叉处共有14条左右的梯架或桥架需要穿越水空仓。为了解决空间处理问题，在电仓的叉口附近，提前把电仓做大成喇叭口，把电桥架由竖向的多层逐个转为平铺的一层，然后在水空仓的最高处穿越回到另一条的电仓内，再由平铺转为多层敷设，解决了难题。

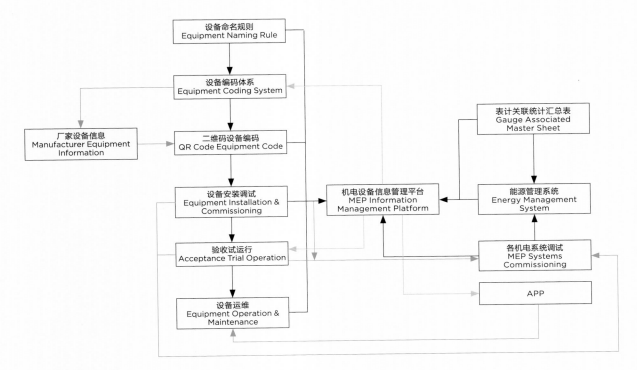

机电设备管理平台流程示意图
MEP Management Platform Flow

7.3.4 Ventilation and AC Design

7.3.4.1 General

1. AC System

Centralized water-cooled central AC system is adopted for air conditioning in summer in the main public areas of T2 and GTC parking building. Direct evaporation type small air-cooled screw conditioner is used as cold source in the substation of the GTC, split air conditioners are used in the scattered small offices on F1-F3 and TIC sub-control center (monitoring hall), and intelligent air-cooled multi-split AC system is provided for some remote offices. Except for rooms with special functions (such as the boarding bridge, VIP area, first class/business class lounge, international hourly charged hotel, etc.), AC and heating is not provided in other areas in winter. Intelligent variable-frequency multi-split AC system is adopted in special function rooms, and air-cooled heating pump units are used for winter heating in the first class/business class lounge in the pier. 4 chilled water systems and 2 refrigerator rooms are provided for T2 and the GTC, which are respectively located on B2 on the east and west sides outside the GTC. The total cooling load is about 119,553 kW (34,000 RT), plus the refrigerating capacity of the small refrigerator operating at night, the total installed cooling capacity is 35,680 RT. The unit cold index of the building is 175W/m².

Energy-saving measures: all-air systems are used for large spaces such as check-in hall, passenger departure hall, arrival hall and baggage claim hall. During the cooling period, the fresh air volume is determined based on the outdoor and indoor enthalpy values. During the transition season, natural ventilation is adopted as far as possible, and the electric smoke exhaust window is opened based on the difference of indoor and outdoor pressures to exhaust the hot air.

Exhaust air heat recovery technology: heat pump driven heat recovery HVF units are used in functional rooms such as offices, VIP areas, first class/business class lounge and hourly charged hotels. When the fresh air flow of the system is ≥3000cmH and the heat-recovered exhaust air rate is ≥70%, the cooling energy of exhaust air is extracted for energy saving.

Operation strategy of AC loads: first, the refrigeration system is composed of both large and small units, which can be flexibly adjusted in response to load changes to prevent waste. Secondly, with cold source group control system, the number of on/off water chillers and corresponding water pumps and cooling towers can be controlled based on the refrigerating capacity. In addition, variable flow control is adopted for the chilled water circulation system, and variable-frequency speed regulation and quantity control is adopted for secondary circulating water pumps.

2. Other Systems

(1) Ventilation system. Garages, MEP rooms, washrooms, baggage storage, smoking rooms, utility tunnels and other are equipped with peacetime ventilation system.
(2) Smoke control and exhaustion system. The smoke control and exhaust system of the building is planned according to the current national fire protection code and performance requirements of China.
(3) Wartime ventilation system for civil air-defense. Civil air-defense shelters are planned on B1 and B2 of the GTC and parking building together with wartime ventilation system.

3. AC Terminals

The maximum air flow of the air conditioners in this project is 130,000m³/h. Variable-frequency air conditioners are used where air flow ≥ 30,000m³/h, and packaged air conditioners are used where air flow ≥ 50,000m³/h. Fresh air conditioners are equipped with primary filter plus electrostatic precipitator; and air conditioners with return air function are equipped with primary filter G4 + medium filter F7.

The maximum air flow of the fan in this project is 155,000m³/h. In the baggage storage, both upper and lower air exhaust systems are adopted, with the lower one closed in case of fire. The fan coil plus air exhaust system is used in small ELV rooms.

7.3.4.2 Design Highlights

1. The project covers a large area thus is equipped with four refrigeration systems. This enables easier hydraulic balance, safe use, fast and easy control and commissioning of the systems.
2. The refrigeration system is composed of both large and small units, which can be flexibly adjusted in response to load change.
3. Concrete basins interconnected by 4 communicating pipes with a diameter of 1000mm are adopted for the cooling towers. The basin depth is 1.5m at the communicating pipe, and 1m at other positions, which well balances the water systems between the cooling towers.
4. To balance the water routes, static balance valve and dynamic differential pressure balance valve are equipped for each terminal air conditioner and branch of fan coil group, balancing the resistance while guaranteeing the safety and reliability of system operation.
5. For rooms such as VIP rooms, first class/business class lounges and hourly charged hotels requiring AC throughout the year, air-cooled intelligent multi-split cooling/heating AC is used for cooling in summer and heating in winter. The system is flexible in control and only requires low operation cost.
6. For washrooms, both lower and upper air outlets are provided, achieving better effect than the approach of providing just upper air outlets.
7. The AC water pipes of the airport refrigeration room enter the terminal through the utility tunnel, and then extend through AC water risers to various areas, guaranteeing both easy maintenance and the required interior clear height of the terminal area.
8. Air intake in the basement garage of the GTC is naturally supplemented via the natural air intake shaft, patio and driveway, and the air intake in the surface garage is naturally supplemented via nearby railings and patios leading to the outside. The ventilation systems are combined with the smoke exhaust system, using shaft ventilation without any horizontal air duct. This improves the clear height of the use space and saves investment.
9. To realize rational airflow organization and aesthetic building, the treated AC air is delivered indoors through side air supply outlets, downward swirl diffusers or ventilation pavilion.
10. Two sets of cooling systems are equipped for

the transformer substation: when the outdoor temperature is 31℃ and the indoor temperature is 35℃, ventilation cooling is adopted; when the outdoor temperature is >31℃ or the indoor temperature is >35℃, AC cooling is used, so as to save operation cost.

11. Considering that some areas of the airport (such as some offices and electrical rooms) need to be used all day long while the cooling capacity is relatively smaller, a set of small AC water pipe system is specially designed, with a small refrigerator for each refrigeration system dedicated for night operation. The design gives detailed consideration to safe and reliable operation of the airport at night.

7.3.5 Building Intelligent System

7.3.5.1 Design Contents
Building intelligent system mainly includes building services management system, energy efficiency management system, MEP equipment information management platform, as well as supporting systems like equipment management network wiring system, computer network system, power distribution, lightning protection and grounding works. ELV, information and security systems for civil aviation purposes are not included.

7.3.5.2 System Composition
1. Functions and Composition
The building services management system mainly monitor and manage AC, plumbing and drainage, lighting, elevator, escalator, APM and other MEP equipment, and can further realize coordinated control between various systems and the flight information through integration. Its subsystems include building services monitoring system, cold source group control system, intelligent lighting control system, dedicated equipment monitoring system, and building services integration management system.

The energy efficiency management system automatically monitors the implementation of the power systems and makes statistics of the water and electricity consumed by the users, based on which a panoramic database and an energy consumption evaluation system are established. Its subsystems include automatic power monitoring system, automatic power metering system, EPS power monitoring system and energy efficiency management platform.

An MEP equipment information management platform is established to define unified codes for main MEP equipment and use QR codes as a display carrier for information sharing and resource integration between various MEP systems and connection between building services management and energy efficiency management system. This can facilitate tracking and correlation of relevant information of the MEP system, and improve the safety and maintainability of equipment.
To achieve the network communications of the above two systems, a set of dedicated computer network and wiring system for building services management is established.

2. Control Rooms
The general control center of building intelligent systems is placed in the equipment room of the MEP management center on F2 in the southwest area of the main building of T2, in which the main servers of all systems are located. Based on building zoning and management demands, 19 MEP management rooms are provided, some of which also serve as sub-control rooms.

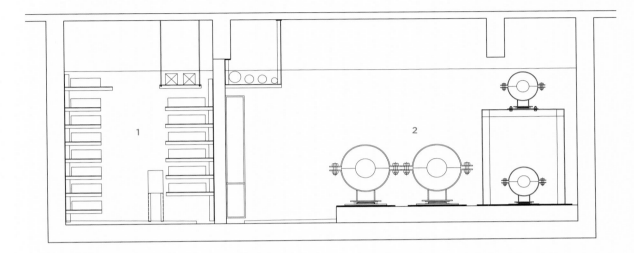

管廊剖面示意
Schematic Section of Utility Tunnel

1.电气管廊
2.空调、水专业管廊
3.电气桥架穿越空调、水专业管廊时,电气管廊平面的喇叭口形状处理

1. Electrical Utility Tunnel
2. AC & Plumbing Utility Tunnel
3. Bellmouth in the Electrical Utility Tunnel Plan When Electrical Cable Tray Crosses AC and Plumbing Utility Tunnel

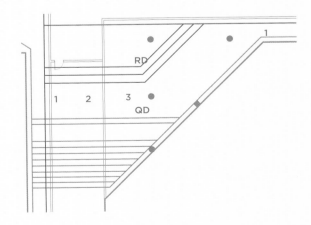

电气桥架跨越空调、给排水管廊典型断面做法示意图
Typical Section of Electrical Cable Tray Crossing AC and Plumbing Utility Tunnel

7.3.5.3 Building Services Management System

1. Building Services Monitoring

This system automatically monitors or controls ordinary AC ventilation, plumbing and drainage, sewage, HVF units, solar hot water, rainwater recycling, electric sunshade louvers, pneumatic smoke exhaust windows, and automatic doors.

The system consists of servers, workstations, network controllers, field controllers, various sensors and actuators, control layer/management layer networks, operating system software and application software, etc., and is integrated with the building services management system. The main server using dual-host redundancy technology is provided in the equipment room of the MEP management cente. Management workstations are provided in the equipment rooms of MEP management center, MEP management rooms serving concurrently as sub-control rooms, and TOC monitoring center; system network controllers (NC) and field direct digital controllers (DDC) are provided in various MEP management rooms or terminal equipment rooms.

The system monitors a total of 11,555 points with its communication network divided into management layer and control layer. The management layer network means the equipment management network, which is a gigabit backbone Ethernet in client/server data processing mode and supporting TCP/IP communication protocol. The control layer network connects various network controllers and field controllers via BACnet bus, allowing for point-to-point communication between controllers.

2. Intelligent Lighting Control System

The intelligent lighting control system controls and manages the lighting in public areas such as passenger check-in hall, security check hall, departure hall, arrival hall, baggage claim hall, VIP room, washrooms and other areas as well as walkways in office areas. The system consists of management servers, workstations, gateways, clock managers, power modules, programmable switch controllers, programmable dimming controllers, programmable control panels, LCD touch screens, input modules, illuminance detectors, human body sensors and management software. The main server using dual hot standby redundancy technology is placed in the equipment room of the MEP management center on F2 of the main building of T2.

3. Building Services Integration Management System

The building services integration management system realizes central integration of the building services monitoring system, cold source group control system, intelligent lighting control system, dedicated equipment monitoring system, and energy efficiency management system of T2 via switched Ethernet, and generates a comprehensive database required for building operation management, so as to centrally manage all global events, and realize information sharing among subsystems and centralized equipment monitoring, alarm management and coordinated control.

7.3.5.4 Energy Efficiency Management System

1. Power Monitoring System

(1) The automatic power monitoring system automatically monitors the HV/LV power distribution systems, transformers, DC panels of T2 (the HV system includes protection and control), and also monitors the door switch state of the substations. The system data is uploaded to the energy efficiency management software platform for analysis. The system is in a three-level network structure of bay layer, station layer and network layer.

(2) Functions: ① automatic monitoring, relay protection and measurement of 10kV HV distribution system; ② automatic monitoring and measurement of power equipment such as 0.4kV LV distribution system, transformers and DC panels; ③ automatic acquisition, analysis and real-time monitoring of the power system operating parameters to timely eliminate potential risks and realize centralized management; ④ monitoring of the door switch state of the power transformation and distribution rooms; ⑤ switch on/off state and fault state signal input terminals of various instruments; ⑥ bidirectional watt-hour measurement, system reading and monitoring of the intelligent instruments at the contact switches; ⑦ over-current "pre-alarming" and "alarming" of intelligent instruments at transformer incoming switches.

2. Automatic Metering System

The automatic metering system remotely acquires the water and electricity consumption data of the rental retail stores, airline facilities, airport occupants' facilities, public washrooms and general meters on some floors of T2. The system data are also uploaded to the energy efficiency management software platform for analysis. The system consists of management server, workstation, communication manager, watt-hour digital meter of bus communication, digital water meter of bus communication, etc. The data of the automatic metering system are preliminarily analyzed by the system management software to generate a report for daily management and charging. The system data are also uploaded to the energy efficiency management system for analysis in combination with the power monitoring system data in the energy efficiency management system platform to finally generate the energy consumption data of the entire terminal.

3. EPS Power Monitoring System

The EPS power monitoring system acquires the operation data of EPS equipment in the power distribution room of the T2 from a standard communication interface, and uploads the data to the energy efficiency management system platform to realize remote monitoring and management. The system is in a two-layer structure. The upper layer is the communication manager located in the MEP management rooms in various zones, serving also the automatic metering system. It is connected upward to the equipment management network for communication with the server of the energy efficiency management system and downward for direct communication with EPS in various power distribution rooms by means of MODBUS bus. This system does not have independent background software. The EPS operation data is directly uploaded to the energy efficiency management system for unified monitoring and analysis using the platform software.

4. Energy Efficiency Management Platform

In addition to the power monitoring and automatic metering systems, the design further acquires and analyzes the consumption of water and electricity in T2, and integrates the management modules. It uses a universal data model to establish a panoramic database, and achieves real-time monitoring, statistical analysis and predictive analysis of the energy consumption system and environmental data through the expert system.

7.3.5.5 MEP Equipment Information Management Platform

1. To strengthen the information management of MEP equipment (systems) in T2, and track and regulate the entire cycle information, i.e. design, production, installation, commissioning, acceptance, use and maintenance via an integrated platform, the Project adopts unified coding technology for its MEP rooms and equipment, establishing a platform system for MEP information management.

2. Construction Contents

(1) A special naming and coding mechanism is established for classified management and unique naming and coding of MEP equipment in T2. The coding also covers specific room and location numbers.
(2) QR code management technology is comprehensively applied to the MEP equipment for listing management and inquiry of information on additional equipment.
(3) Overall control of equipment life cycle is achieved to realize all-process effective tracking and information consistency of equipment design, production, installation and use, facilitate the tracking of relevant information, and improving the safety and maintainability of the equipment.
(4) Advanced, efficient and expandable unified information platform: a unified management platform and database is established, and is interfaced with all MEP management systems (building services management system and energy consumption management system) platforms in T2 to realize data sharing.

7.3.5.6 Design Highlights

1. Decentralized control and centralized management. In view of building zoning and management demands, 16 equipment management rooms are provided in T2, some of which are also used as sub-control rooms, to serve the management mechanism of long-span general control and sub-control of T2.
2. Dedicated wire and network and high-speed communication. To satisfy the network communication of various building ELV systems (except the automatic fire alarm system), a set of dedicated computer network and wiring system for building services management is provided. In addition, single mode fiber is used as the transmission medium for the main cable with double 10G switches as the core network, which solves the difficulty of long-span communication in T2.
3. Comprehensive supervision and high integration. The design comprehensively regulate AC, plumbing and drainage, lighting, elevators, escalators, APM, pneumatic smoke exhaust windows, automatic doors, electric sunshade louvers, and other MEP equipment in T2 on the building services management platform, and realizes coordinated control of all systems via further integration to upgrade the management level of T2.
4. Optimized energy efficiency and intelligent management. The energy efficiency management system makes statistics on the power system, terminal water and power consumption, based on which a panoramic database and an energy consumption evaluation system are established to analyze the energy consumption of the building and propose the technical and management measures for energy saving and consumption reduction. Additionally, to facilitate the management of MEP equipment in T2, an MEP equipment information management platform is established to provide uniform codes for all major MEP equipment, with QR codes as display carriers to realize the uniqueness of the equipment supervised by the building services management and energy efficiency management system and also facilitate the data interconnection of the two.
5. Adjustment based on local conditions and colorful dimming. Based on the characteristics of departure/arrival halls, check-in islands, VIP lounge, BOH office area and other places in T2, various technologies such as 0-10V dimming, DMX512, DALI dimming and switch control are adopted. Among them, DMX512 stage lighting control technology in combination with WRGB floodlights above the check-in islands can realize colorful dimming effect and offer pleasant experience to passengers.

7.3.6 Design of Utility Tunnel

7.3.6.1 Design Highlights

1. Characteristics of route planning. The route planning is based on building layout and characteristics in combination with the equipment room locations of various specialties, which not only ensures the safety and easy installation and maintenance of pipelines, but also makes maximal use of the utility tunnel area.
2. Different compartments for plumbing and electrical pipelines. The plumbing pipelines and electrical pipelines are arranged in different compartments of the utility tunnel to guarantee fire protection and avoid any trouble caused by their interwoven layout to construction and management.
3. Section design. The sectional dimensions are determined in consideration of the laying of plumbing and electrical pipelines, as well as construction space, maintenance space and fire prevention and evacuation.

7.3.6.2 Design Challenges

The major design challenge is the crossing of electrical pipelines through the empty plumbing compartment. There are more pipelines and more layers of trays in the electrical pipeline compartment, with about 14 ladder or cable trays that need to pass through the empty water compartment at the intersection. To solve this, the design makes the part of electrical pipeline compartment near the crossing into a big bell mouth in advance to change the electrical cable tray gradually from vertical multi-layer to horizontal one layer layout, then has the electrical pipeline crosses back to the electrical compartment of another tray via the highest point of the empty plumbing pipeline compartment, where the horizontal layer is changed back to vertical multi-layer layout.

7.4 市政工程设计

7.4.1 市政道路及停车场工程设计

7.4.1.1 概述

二号航站楼陆侧交通系统主要包括北进场隧道、东环路、出港高架桥、西环路、南往南高架桥等28条道路，交通中心南面北进场隧道东西两侧设置了两个地面停车场，地面停车场总面积53089m²，共有私家车位1528个。

7.4.1.2 市政道路工程设计

1. 各类车辆良好流线的设置

（1）南向入口交通组织：①私车到达接客到场：从南进场→东连接楼→东停车场、停车楼、西停车场。②私车到达接客离场：东停车场、停车楼、西停车场→西连接楼→到南出口。③（的士出发送客）从南进场→东连接楼→二号楼出发高架→西连接楼→到南出口。④（大巴到达接客离场）从南进场→东连接楼→二号楼出发高架→西连接楼→到南出口。

（2）北向入口交通组织：①私车到达接客到场：从北进场隧道→东停车场、停车楼、西停车场。②私车到达接客离场：东停车场、停车楼、西停车场→出北进场隧道。③的士出发送客离场：从北进场隧道→北停车场→二号楼出发高架二号楼出发高架→北停车场→出北进场隧道。④大巴到达接客到场：从北进场隧道→北停车场→东连接楼→二号楼出发高架。⑤大巴到达接客离场：二号楼出发高架→西连接楼→出北进场隧道。

2. 典型道路横断面设计

出港高架桥（K0+780~K0+980）段设置大巴隧道和的士隧道下穿中央广场，中央广场处这两条隧道下方为北进场隧道，11.25m层为10车道的出港高架桥，横断面布置如图：出港高架桥道路断面设计。

7.4.2 市政桥梁工程设计

7.4.2.1 桥梁概况及规模

广州白云国际机场扩建工程子项市政桥梁工程共包括主线桥一座，匝道桥五座，分别为出港高架桥、交通中心私家车坡道、南往南匝道、北进场东匝道、北进场东匝道、东三匝道和西三匝道。

7.4.2.2 市政道路工程设计

除满足基本的交通功能而有自身超宽、异形等特点外，还具有以下特点：

1. 二号航站楼出港高架结构特点。本次出港高架主要是为满足二号航站楼出发层使用，起终点为标准单向三车道，桥宽12.25m。楼前段为满足送客停车要求，桥宽达到了52.45m，纵向跨度36m，横向设置三个支座，最大跨径23m，桥宽远超于跨度，空间受力效应显著。不同桥宽段通过中间约120m弧线段过渡，桥宽从12.25m变化到52.45m，曲线半径最小45m，属于超异形结构。

2. 跨线桥结构特点。南往南跨线桥位于一号和二号航站楼之间，主要为满足车辆掉头和进入贵宾厅的功能，共设置五条匝道。受机场东环路、机场西环路、北进场路东线北进场路西线、地铁和城轨等影响，直线段最大跨径为59m，跨路曲线段最大跨径50m，具有结构跨度大，平面曲线半径小的特点。

3. 桥梁结构在复杂综合体中设计、施工衔接。出港高架及跨线桥与二号航站楼其他子项、地铁、城轨等都存在一定的衔接和交叉，存在大量共柱、共基础等问题，设计过程中与各方设计部门进行沟通衔接，并配合业主协调施工界面和施工进出场顺序。

7.4.3 市政隧道工程设计

7.4.3.1 概况

市政隧道工程项目范围共包括三条隧道，分别是北进场隧道、到港巴士隧道和到港的士隧道。

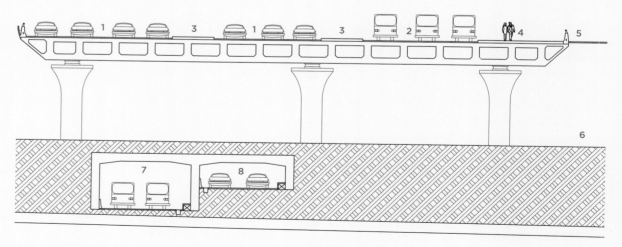

1. 出发层小型车车道
2. 出发层大巴车道
3. 安全岛
4. 人行道
5. 门斗连桥
6. 0m层人行道
7. 大巴隧道
8. 的士隧道
9. 北进场隧道

1. Departure Floor Car Lane
2. Departure Floor Bus Lane
3. Traffic Island
4. Sidewalk
5. Bridge-foyer Connection
6. Sidewalk at Level 0 m
7. Bus Tunnel
8. Taxi Tunnel
9. North Approach Tunnel

出港高架桥道路断面设计
Departure viaduct section design

7.4.3.2 设计特点

跟常规设计有所不同，大部分隧道节段的中墙处设置有航站楼、交通中心、市政桥梁等结构的桩基础，因此本项目隧道考虑砂层与隧道桩基共同受力，采用不同的换算刚度模拟隧道基底支撑全部的上部荷载。

到港巴士、的士隧道在地下负一层位置分别在东侧、西侧与城轨和地铁的站厅顶板相交。其中部分节段采用与城轨、地铁站厅共楼板和支撑柱的方式合建。除需考虑常规的恒载、车辆竖向荷载、土的覆重及侧向土推力外，还需考虑车辆制动力引起的水平推力对下部结构立柱的影响。

二号航站楼隧道下穿机坪段市政隧道荷载设计活荷载采用机坪通行的最大飞机"-B747-400"荷载，利用刚性地基梁原理进行计算，隧道结构采用单箱双室矩形结构，通过结构自重和墙趾的覆土压重进行结构抗浮。隧道地基采用复合地基。

7.4.3.3 隧道排水及消防

1. 隧道排水工程。根据《室外排水设计规范》（GB50014-2006）要求，隧道排水设计标准按20~30年一遇暴雨重现期设计。本项目由于项目性质重要，位置特殊，设计时按50年一遇暴雨重现期设计，100年一遇暴雨强度进行校核。

2. 隧道消防给水设计。北进场路隧道属于二类隧道，大巴隧道及的士隧道属于四类隧道。根据《建规》规定，北进场路隧道须设置室内消火栓系统，大巴隧道和的士隧道可不设置消防水系统。隧道内消火栓水量为20L/s，隧道洞口消防水量为10L/s。火灾延续时间按3小时考虑。

7.4.3.4 隧道电气

北进场隧道属于二类隧道，大巴隧道及的士隧道属于四类隧道。设备安装内容主要包含配电系统、电气照明系统、建筑物防雷与接地、消防及弱电系统。北进场隧道照明分为以下6种：入口段照明、过渡段照明、中间段照明、出口段照明、基本照明及疏散照明。

消防报警系统选用智能型区域报警系统，对隧道内的火灾信号和消防设备进行监视及控制，并预留与上一级控制中心信息共享的接口。报警系统系统组成有：火灾自动报警系统、消防联动控制系统、火灾应急广播系统、消防专用电话系统、应急照明控制及消防系统接地。

7.4.3.5 隧道通风

1. 设计原则。（1）正常交通情况下，为司乘人员、维修人员提供合理的通风卫生标准，为安全行车提供良好的空气清晰视度；（2）火灾事故情况下，通风系统应能够控制烟雾和热量的扩散，而且为逗留在隧道内的司乘人员、消防人员提供一定的新风量，以利于安全疏散和灭火扑救；（3）在确保通风可靠性及节能运行、节约工程投资的前提下优选适当的通风方式；（4）通风设备选用原则是：安全可靠，技术先进。高效、节能的设备按近、远期合理配置；（5）控制对工程环境质量（环境空气质量、噪声）的影响，注重环境保护措施。

2. 设计参数叙述。（1）隧道内通风卫生标准。CO允许浓度正常工况：150ppm；阻塞工况：300ppm；（2）隧道噪声标准。行车隧道内：<90dB（A）；环境噪声标准：昼间等效声级60dB（A），夜间等效声级55dB（A）；（3）尾气排放标准。CO基准排风量：0.01m³/辆·km；烟雾基准排风量：2.5m³/辆·km。

3. 风机布置。根据计算，单向隧道内设置3组、每组2台射流风机(单向)，每组射流风机的净距不小于1倍风机叶轮直径，便可满足隧道通风要求，考虑到隧道局部沉板的影响，射流风机安装在隧道顶板拱形顶部区域，最后分别于隧道+290、+472.177及+678.177各设置1组射流风机，并且在隧道靠近出口处(长隧道需间隔200m~300m)设置一套VI-CO分析仪，检测隧道内CO与烟雾浓度，控制风机开停。

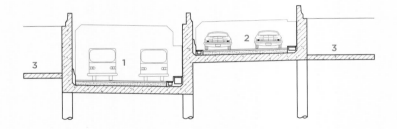

市政隧道敞口段剖面图
Opening Segment Section of Utility Tunnel

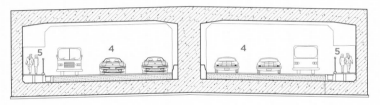

市政隧道机坪段标准剖面图
Apron Segment Typical Section of Utility Tunnel

1.巴士隧道
2.的士隧道
3.地铁站厅顶板
4.车行道
5.人行道

1. Bus Tunnel
2. Taxi Tunnel
3. Top Plate of Metro Station Concourse
4. Vehicular Lane
5. Pedestrian Lane

7.4 UTILITIES

7.4.1 Municipal Road and Parking Lot

7.4.1.1 General
The landside transportation system of T2 has 28 roads including the north arrival tunnel, the east ring road, the departure viaduct, the west ring road, and the south-to-south viaduct. Two ground parking lots are provided on the east and west of the north arrival tunnel on the south of GTC. With a total area of 53,089m², the parking lots offers 1,528 private parking spaces.

7.4.1.2 Municipal Road
1. Smooth Vehicular Circulations
 (1) Traffic organization at the south entrance: ①.private cars arrive at the drop-off areas: the south arrival tunnel → the east connecting building → the east parking lot, the parking building or the west parking lot .②.Private cars arrive at the pick-up areas: the east parking lot, the parking building or the west parking lot → the west connecting building → the south exit. ③. Taxis arrive at the pick-up areas: the south arrival tunnel → the east connecting building → the departure viaduct from T2 → the west connecting building → the south exit. ④.Coaches arrive at the pick-up areas: the south arrival tunnel → the east connecting building → the departure viaduct from T2 → the west connecting building → the south exit.
 (2) Traffic organization at the north entrance: ①. private cars arrive at the drop-off areas: the north arrival tunnel → the east parking lot, the parking building or the west parking lot. ②.Private cars arrive at the pick-up areas: the east parking lot, the parking building or the west parking lot → the north arrival tunnel. ③. Taxis arrive at the pick-up areas: the north arrival tunnel → the north parking lot → the departure viaduct from T2 → the north parking lot → the north arrival tunnel. ④.Coaches arrive at the drop-off areas: the north arrival tunnel → the north parking lot → the east connecting building → the departure viaduct from T2. ⑤.Coaches arrive at the pick-up areas: the departure viaduct from T2 → the west connecting building → the north arrival tunnel.

2. Typical Road Cross Section
In the departure viaduct (K0+780-K0+980) section, a tunnel for coaches and one for taxis pass through the central square, and beneath the tunnels is the north arrival tunnel. The 10-lane departure viaduct is at 11.25m level. The cross section of the viaduct is as follows.

7.4.2 Municipal bridge

7.4.2.1 Bridge General and Size
As part of the Airport Extension Project, the municipal bridge subproject includes a main bridge and five ramp ones, namely the departure viaduct, the ramp for private cars from GTC, the south-to-south ramp, the east/west ramp of the north arrival tunnel, the third east ramp and the third west ramp.

7.4.2.2 Municipal Road
The roads are beyond conventional width and geometry and presents the following features:

1. The structural characteristics of the departure viaduct from T2. The departure viaduct is mainly built to meet the needs for vehicles from the departure floor of T2. The 12.25m wide bridge starts and ends with three standard one-way lanes. The part of the bridge in front of the Terminal is 52.45m wide to meet the requirements for parking. Horizontally, the bridge has three supports, and the maximum span is 23m, which is remarkably smaller than the bridge width, boasting significant spatial force effect. With a curved section of 120m long, the width of the bridge soars from 12.25m to 52.45m. The bridge has a geometry significantly diverting from the conventional ones as the curve radius is at least 45m.

2. Structural characteristics of the overpass. The south-to-south overpass located between T1 and T2 is designed for vehicles turning around and entering into the VIP area. A total of five ramps are developed. Affected by the east and west ring roads, the east and west branches of the north arrival road, the metro line and the intercity railway line, the overpass spans a maximum of 59m in the straight section and 50m in the curved section, boasting large structural span and small radius of plane curve.

3. The bridge structure is designed, constructed and connected within a complex. The departure viaduct and overpass are connected to and overlapped with other subprojects of T2, the metro line and intercity railway line, as they share a large number of columns and foundations. Interdisciplinary communication was conducted to assist the client in coordinating the construction interfaces and sequences.

7.4.3 Municipal Tunnel

7.4.3.1 General
The municipal tunnel project involves three tunnels, namely the north approach tunnel, arrival bus tunnel and arrival taxi tunnel.

7.4.3.2 Design Highlights
Different from the conventional design, most of the tunnel sections have the pile foundations of the terminal building, GTC and municipal bridges. Therefore, sand layers and the tunnel pile foundations are used to joint support the upper loads, which is simulated with different conversion stiffness.

The arrival bus and taxi tunnels intersect with the roofs of the station concourses of the intercity railway and metro lines at B1 on the east and west. Some of the tunnel sections share roofs and supporting columns with the station concourses. In addition to the conventional dead load, vertical load of vehicles, soil weight and lateral earth pressure, consideration is also given to the impact of the horizontal thrust caused by the vehicle braking force on the lower structural columns.

The live load on the section of the T2 municipal tunnel passing through under the aprons is the maximum load on the rigid foundation beams when the aircraft "-B747-400" travels on the aprons. The tunnel consisting of double-chamber rectangular boxes resists floating with its own structural weight and the soil weight of the wall toe. A composite foundation is provided for the tunnel.

7.4.3.3 Tunnel Drainage and Fire Protection
1. Tunnel drainage. According to the *Outdoor Drainage Design Code (GB50014-2006)*, the tunnel drainage should be designed to accommodate rainstorms recurring every 20 to 30 years. In view of the significance and special location of the project, the drainage design of the T2 follows the requirements for a recurrence interval of 50 years and the review requirements for 100 years.

2. Water supply for tunnel fire protection. The north arrival tunnel is a Class 2 tunnel, and the tunnels for coaches and taxis are Class 4 ones. According to the Code, an indoor fire hydrant system is mandatory for the north arrival tunnel but optional for the tunnels for coaches and taxis. The water supply capacity of the fire hydrant is 20L/s in the tunnel and 10L/s at the tunnel entrance, which is based on a 3-hour fire duration.

7.4.3.4 Tunnel Electrical
The north arrival tunnel is a Class 2 tunnel, and the tunnels for coaches and taxis are Class 4 ones. Installed facilities mainly include power distribution systems, electrical lighting systems, lightning protection and grounding systems, fire protection and ELV systems. There are six types of lighting in the north arrival tunnel: lighting in the entrance, transition, middle and exit sections, general lighting and evacuation lighting.

Intelligent local fire alarm systems are adopted to monitor and control the fire signals and fire facilities in the tunnel, and interfaces to share information with the control center are reserved. The systems consist of an automatic fire alarm system, a linked control system, an emergency broadcast system, a special telephone system, an emergency lighting control and fire protection grounding system.

7.4.3.5 Tunnel Ventilation
1. Design principles. (1) Under normal traffic conditions, to ensure good ventilation and sanitation for drivers, passengers and maintenance personnel, and great visibility for safe driving under normal traffic conditions. (2) In case of a fire, to prevent the smoke and heat from diffusion and provide a certain amount of fresh air for the passengers and firefighters in the tunnel to facilitate evacuation and firefighting. (3) To provide a proper ventilation solution to ensure reliable ventilation, energy-saving and cost-effective operation. (4) To select safe, reliable and advanced ventilation equipment. Efficient and energy-saving equipment are to be reasonably provided in the near and long terms. (5) To take environmental protection measures for minimal impact on the environment (air quality, noise).

2. Design parameters. (1) Standards for ventilation in tunnels. Permitted CO concentration under normal working conditions: 150ppm; that during traffic jams: 300ppm. (2) Standards for noise control. Inside the tunnel: <90dB (A); ambient noise standard: Equivalent Continuous Sound Pressure Level in daytime: 60dB (A), that at night: 55dB (A). (3) Standard for exhaust emission. Base CO exhaust: $0.01m^3$/km; base smoke exhaust: $2.5m^3$/km.

3. Fan layout. The calculation results show that, the requirements for ventilation can be met by installing three groups of one-way jet fans (each group consists of two fans), with the distance between two groups being no less than 1 time the diameter of the fan impeller. As some parts of the ceiling of the tunnel are sunken, the jet fans are installed in the arched areas of the tunnel ceiling. The three groups of fans are installed respectively in the positions of +290, +472.177 and +678.177 of the tunnel. Furthermore, a VI-CO analyzer is installed near the tunnel exit (long tunnels need to be separated by a distance of 200 to 300m) to detect the concentration of CO and smoke in the tunnel so as to turn on or off the fans.

7.4.4 地下综合管廊

7.4.4.1 概述
根据二号航站楼的总体规划,由于二号航站楼和机坪切断了现状机场南、北两区的联系,为保证二号航站楼建成后,南北两区的管线联系继续保持畅通,根据《广州白云机场扩建工程可行性研究报告》规划及与白云机场管线权属单位讨论沟通,最终设计提出沿机场北进场路两侧修建两条大型综合管廊,管廊北端通过两条小型综合管廊与一期已建综合衔接,管廊南端与现状东三、西三指廊附近现状管线衔接。管廊主要用于收集现状北进场路管线及进出二号航站楼新建管线,以及预留远期南北两侧须新增的联络管线。

7.4.4.2 设计原则
1. 分舱原则。(1)管线分舱以管线自身敷设环境要求为基础,在满足管线功能要求的条件下可根据规划管线数量、管径等条件合理同舱;(2)电力与通信管线可兼容于同一舱室,但需注意电磁感应干扰的问题;(3)110KV电力管道由于使用需求,单独一个舱室;(4)通信管道可与供水同设一个舱室。

2. 内部管线排列原则。为集约化利用管廊空间,多层布设。自地表向下的排列顺序为:消防管、电力管、通信管和给水管。

3. 断面设计原则。(1)综合管廊断面充分考虑管线安装空间和必需的检修通行空间,同时满足管径最大的管道和管件具备更换的操作空间;(2)综合管廊断面应兼顾各种设施、阀门、控制设备及设备更换的空间;(3)综合管廊断面型式的确定,要考虑到综合管廊的施工方法及纳入的管线数量。本项目施工采用明挖为主,因此综合管廊的断面形式采用矩形断面。

7.4.4.3 综合管廊总体设计
市政综合管廊管廊包括给水管、压力污水管、10KV中压电缆、电信电缆、航空电信电缆。室外综合管廊沿二号航站楼中轴线东西对称布置,呈南北走向,总长度约2.2km,宽度为6.4~7.1m;室内综合管廊在二号航站楼下方,呈网状布置,与建筑布局一致,总长度约7.4km,宽度为6.0~15m。根据管廊工艺、通风、消防、排水等需要,管廊每隔一定距离设置防火分隔、投料口、通风口、逃生口、集水井等附属设施。

7.4.4.4 综合管廊设计特点及难点
1. 第一次将市政综合管廊设计理念引入大型综合交通枢纽建筑中来,为同类建筑室外管线综合设计提供了案例参考。较全国大力推广建设综合管廊时间提早了3年。
2. 综合管廊的设置解决了机场南北向市政管线没有埋地敷设的条件,为机场大量现状管线提供了管道敷设空间,同时也为机场后期发展预留了管道敷设空间。
3. 室外综合管廊与楼内综合管廊无缝对接,室外市政管线可以直接接入楼内综合管廊和设备机房。
4. 管线检修维护均在综合管廊内部,减少了对机场地面的开挖,降低了管线开挖时对机场运营的影响,提高了管线的使用寿命。
5. 节约室外敷设管线的用地,提高机场土地使用率。

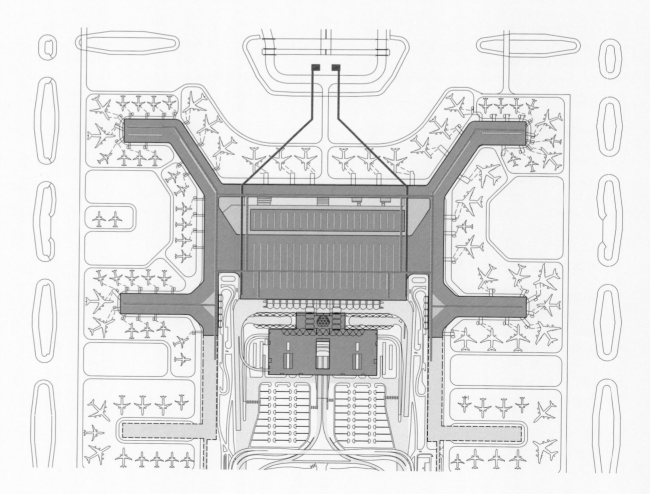

综合管廊总平面布置图
Master Layout of Integrated Utility Tunnel

— 室外综合管廊
Outdoor Integrated Utility Tunnel

— 楼内综合管廊
Indoor Integrated Utility Tunnel

7.4.4 Utility Tunnel

7.4.4.1 General

In its master plan, T2 and the apron cut off the connection between the existing north and south areas of the airport. To link the pipelines in the two areas after the completion of T2, based on the planning specified in the *Feasibility Study Report on Guangzhou Baiyun Airport Expansion Project* and communication with the authorities, two large-sized utility tunnels are constructed along both sides of the north arrival tunnel. The utility tunnels have been connected with the existing pipelines in Phase I through two small-sized utility tunnels in the north, and with those around the existing 3rd east and west piers in the south. The utility tunnels are mainly built to accommodate the existing pipelines along the north arrival road and the new pipelines from and to T2, and to reserve space for the connecting pipelines to be added on both the north and south in the long run.

7.4.4.2 Design Principles

1. Cabins. (1) Cabins may be used to lay the pipelines. On the premise of ensuring the functions of the pipelines, cabins may be used to accommodate pipelines based on the quantity and diameters of the planned pipelines. (2) The power supply and communication pipelines can be placed in the same cabin. However, electromagnetic interference should be avoided. (3) 110kV power supply pipelines are put in individual cabins to meet the requirements for usage. (4) The communication water supply pipelines can be provided in the same cabin.

2. Internal pipeline arrangement. Pipelines are laid on multiple layers for the intensive use of the tunnel space. They are provided from the surface down in the following order: fire hoses, power supply pipes, communication pipes and water supply pipes.

3. Section design. (1) The design of the utility tunnel section provides sufficient space for installation and service, and for the replacement of pipes and pipe fittings with the largest diameter. (2) The section also provide space for various facilities, valves and controls and their replacement. (3) The type of section is determined partly based on the method for the construction of the utility tunnel and the number of pipelines included. As open excavation is mainly adopted in the project, the section is a rectangle.

7.4.4.3 Overall Design

The utility tunnel includes water supply pipes, pressurized sewage pipes, 10KV medium-voltage cables, telecommunication cables, and aeronautical telecommunication cables. The outdoor utility tunnels are provided from north to south symmetrically along the central axis of T2 on the east and west. They are about 2.2 km long in total and 6.4 m to 7.1 m wide. The indoor ones are below T2 and in the form of a mesh. With the layout being consistent with the architectural layout, the tunnels are about 7.4km long in total and 6.0m to 15m wide. According to the requirements for utility tunnel technology, ventilation, fire protection, drainage, etc., auxiliary facilities such as fire compartments, feeding ports, vents, exits and sumps are provided in the tunnels at certain intervals.

7.4.4.4 Design Highlights and Challenges

1. For the first time, the concept of utility tunnel is incorporated into the development of a large-scale comprehensive transportation hub, which may serve as a reference for the design of outdoor pipelines of similar buildings. The project has been completed three years ahead of the promotion of the construction of utility tunnels nationwide.
2. The utility tunnels provide space for laying the existing south-north municipal pipelines and future ones in the airport.
3. The outdoor utility tunnels are seamlessly connected with the indoor ones, and the outdoor municipal pipelines can be directly connected with the utility tunnels and equipment rooms in the building.
4. Pipelines can be maintained and repaired inside the tunnels, which reduces the needs for excavation, hence less impact on the airport operation during pipeline excavation and longer service life of the pipelines.
5. The design saves land for laying outdoor pipelines and improves the airport land use.

7.4.5 市政给水系统设计

7.4.5.1 概述

二号航站楼市政给水系统由室外生活给水管网系统、室外消防加压管网系统、室外绿化浇洒管网系统组成。

7.4.5.2 室外给水系统设计

1. 室外给水管网布置及敷设。二号航站楼及交通中心停车楼生活给水管由机场两条DN500的市政供水主管供应。同时预留地铁及城轨生活及消防用水。预留接口均从管廊内的DN500市政供水主管接出供水。南侧东西两个停车场地块预留DN200给水接口供地块远期开发利用。生活供水管网布置成环状供水，管网沿程布置室外消火栓，室外消火栓间距不大于120m，保护二号航站楼及陆侧建筑；
2. 室外消防加压管网布置。二号航站楼室内外消防系统共用加压系统，系统满足机场室内外消防流量要求，消防加压管网管径采用DN300，加压管沿二号航站楼环状敷设，室内消火栓及自动喷淋系统就近从消防加压环管上引入室内消防管网。

7.4.5.3 室外给水系统设计特点

1. 生活给水供水安全度高。生活给水水源采用机场南北供水站供水，双水源供水有保障；
2. 室内消防系统共用加压管网系统，简化消防控制系统，节省室外消防加压管路；
3. 二号航站楼消防系统与一号航站楼消防系统连接，消防供水安全性高；
4. 消防加压系统考虑室外消防水量，可减少消防车转输室外消防水量的需求。

7.4.6 市政排水系统设计

7.4.6.1 概述

市政排水系统设计主要为室外生活污、废水管道设计和室外雨水管网设计。排水体制为雨、污、废水完全分流制，粪便污水经化粪池处理后排入市政污水管网，营业餐饮废水经隔油池处理后排入市政污水管网。

7.4.6.2 室外污水系统设计

1. 排水管网设计。根据排水现状情况，二期扩建工程污水由于没有重力污水管接入条件，故设计考虑集中在二号航站楼西南侧高架桥下设置集中污水泵站，二号航站楼污水经新建污水泵站收集后，排入1号污水泵站，1号污水泵站再将污水泵送至污水处理厂附件的2号污水泵站，集中送往机场污水处理厂进行处理；
2. 排水管道布置及敷设。二期场内室外污水管网收集建筑物内生活废水及化粪池之后污水管道，陆侧管道沿建筑物周边道路敷设，空侧污水管沿指廊外侧服务车道边敷设。

7.4.6.3 室外雨水系统设计

1. 雨水设计参数。考虑地域气候及项目重要程度，室外暴雨重现期取10年，防洪标准采用50~100年重现期，径流系数取0.85。
2. 峰值雨水量预测。根据场地及建筑布置分析，经过计算，陆侧峰值雨量约为14.87m³/s，空侧峰值雨量约为12.9m³/s。
3. 室外雨水系统设计。根据现状排水情况，二期扩建工程雨水南、北均没有排出接口，唯一可考虑向东西两侧排至空侧机坪雨水渠箱。根据与民航总院沟通，二号航站楼屋面靠空侧一边的雨水直接就近排入空侧服务车道雨水渠。二号航站楼靠陆侧屋面、交通中心屋面及陆侧场地雨水经场内室外雨水管道收集后向西南角1号雨水泵站排放，再通过1号雨水泵站提升至空侧机坪内的雨水渠箱。
4. 室外排水系统设计特点。
（1）完全分流排水体制，满足市政排水环保要求。
（2）大量运用新材料、新工艺，提高工程质量，增加污水管网使用寿命。
（3）应用大型雨水调蓄设施，提高机场应对极端天气能力。

7.4.7 海绵城市在二号航站楼的应用

7.4.7.1 概述

1. 海绵城市
通过"海绵城市"建设，充分保护自然水体和植被，最大限度地减少建设开发对原有生态环境的破坏，有效地削减降雨径流总量和降雨峰值流量，减轻排水压力。

2. 机场水环境系统
（1）易涝性分析。项目范围原场地为预留空地，径流系数不到0.2。机场开发扩建后，场地将建设成为市政道路、飞机跑道、二号航站楼等设施，场地硬质化覆盖率发生了极大变化，雨水径流系数将由现状的0.2提高至0.8，进而造成雨水径流总量及峰值径流量增大、峰值汇流时间缩短，暴雨时极易形成内涝灾害；（2）径流污染分析。机场扩建后将出现大量建筑群落、道路等非点源污染源，势必会造成一定程度的径流污染；（3）径流限制性因素分析。受排水条件限制，机场二期扩建工程地块内无现状雨水管道，二号航站楼陆侧雨水只能经由1号雨水泵站提升排至外江；1号雨水泵站受双孔箱涵限制，排水能力只能扩大至12.8m³/s，且泵站前池容积无法增大。二号航站楼扩建后，原1号雨水泵站排水能力无法满足场地排水需求；（4）对策分析。由于外排流量受限，排水防涝需求、新机场水系补水蓄水量较大，应采用雨水控制与利用设施调节径流峰值、调蓄径流总量等。

7.4.7.2 海绵城市规划设计要求

1. 总体目标。在机场建设过程中应用海绵城市设计理念，提高场地防洪排涝减灾能力、改善生态环境、缓解城市水资源压力。根据排水防涝和"海绵机场"建设目标，结合广州新白云国际机场水系统的特点，确定径流总量控制、径流污染控制、排水防涝控制、水环境保护为主要综合控制目标。
2. 设计策略。根据降雨、土壤等因素，综合考虑水环境、水资源、水生态、水安全等方面的现状问题和建设需求，具体实施策略为：（1）工艺方面。优先选择具有水质、水量等综合作用的生物滞留设施、雨水花园等，并综合考虑性价比和景观效果。加大场地雨水的自然渗透效果，补充地下水。工艺整体侧重于对径流总量的控制，并和排水管网、调蓄池等措施结合，确保排水防涝能力的达标；（2）面源污染控制方面。侧重于水质的控制，将其作为控制面源污染的重要组成部分，重点控制TSS、总氮、总磷等污染指标，并且和截污、清淤等措施结合起来，统筹解决水环境治理问题；（3）景观环境协调方面。室外设置生态停车，结合景观绿化，利用植物爬藤，改善停车环境，提高植物群落水土保持和涵养水源的能力。

7.4.7.3 海绵城市设计要点

1. 适宜技术选择。从延长雨水汇流时间、削减峰值的角度，宜优先选择一些具有调蓄功能的设施；从缓解径流初期污染的角度，宜选择具有净化功能的设施；从雨水资源化利用的角度，宜选择具有储水功能的设施，并辅以植草沟等雨水输送设施。
2. 海绵城市系统构建。利用SWMM模型软件对场地排水进行模拟，通过对地块整体径流量和管道流量模拟，计算末端出水口的动态径流过程，进而分析项目LID设施的效果和作用。
3. 场地低影响开发设计。结合机场特点，本项目选用的LID设施类型主要为下沉式绿地、透水铺装道路、雨水花园、绿色树池等。

7.4.8 市政电气设计

7.4.8.1 照明系统设计

1. 照明工程设计原则。（1）道路照明设计以功能性为主，并考虑一定的景观性。道路照明的质量也应达到国内先进水平，且符合现代都市的发展要求和品位；（2）道路照明除使道路表面满足亮度及照度均匀度要求外，使驾驶人员视觉舒适，并能看清周围环境；（3）绿色照明，选择高效光源及灯具，光源选用高压钠灯及LED灯，灯具效率不低于0.7钠灯/0.9 LED灯，防护等级不低于IP65，灯具、灯杆造型与机场整体环境协调、美观。
2. 照明工程设计标准。因本项目道路、基础设施用地、广场等相结合，拟采用区域照明结合方式进行设计。
3. 道路照明设计方案。地面道路及高架桥布灯形式：（1）本工程因地面道路、桥面道路及停车场等相结合，为了避免灯杆林立的问题，采用高杆灯照明，与周边建筑环境相协调。高杆灯采用5×1000W（初始流明120000Lm）高压钠灯，灯具装高30m；（2）出港高架桥下地面道路：桥底道路采用吸顶灯布置方式，灯具采用60W LED吸顶灯，装于两桥墩中间，灯具纵向间距5米。

7.4.8.2 交通监控及停车场系统设计
1. 交通监控设计。监控点位设置：本工程交通监控共分为车场视频监控、治安及道路视频监控、治安视频卡口监控、交通违法抓拍四部分。
2. 停车场管理系统。（1）停车场分类设置：停车楼、室外停车场、VIP停车场、大巴停车场、出租车上客车场；（2）停车楼及室外停车场系统功能：停车场出入口管理、出入口车牌自动识别、场内二次识别、室内车位引导、反向寻车、会车报警、中央收费、自助收费、人工收费；（3）VIP停车场系统功能：停车场出入口管理、车牌自动识别；（4）大巴停车场系统功能：停车场出入口管理；（5）出租车上客车场系统功能：车场入口管理。

7.4.8.3 市政动力系统设计
1. 本工程调蓄水池水泵用电负荷属二级负荷，为满足供电可靠性要求，由室外两个变电站低压柜各取1路380V电源，一用一备。泵站采用PLC自动控制水泵及格栅机启停，主要设备负荷：潜水泵、排泥泵、格栅机、启闭机、反冲洗设施及其他安装负荷约158kw；计算负荷108.8kw。控制方式为手动模式、遥控模式、自动模式。
2. 技术应用。本工程自控系统设计采用开放式结构体系的自动化系统，将系统与设备有机结合在一起用于监控生产。将信息流扩展到整个生产过程，从而实现过程控制数据与信息方便可靠地在PLC与外部设备之间交换。信号监测：雨水调蓄水池水位计及泥位计，将所有检测参数和设备运行状态实时传送至中心控制室。

7.4.5 Municipal Water Supply System

7.4.5.1 General
The municipal water supply system of T2 consists of an outdoor domestic water supply network, an outdoor fire protection pressurized network, and an outdoor greening sprinkling network.

7.4.5.2 Outdoor Water Supply System
1. Arrangement and laying of outdoor water supply network. The domestic water is supplied to T2 and the parking building of GTC by the two main DN500 municipal water supply pipes. At the same time, there are reserved interfaces to connect pipes for domestic and fire protection water for the metro and intercity railway. The reserved interfaces are all connected to the DN500 municipal water supply pipes in the utility tunnels. DN200 water supply interfaces are reserved for the long-term development and utilization of the east and west parking lots in the south. The domestic water supply pipe network forms a ring, and outdoor fire hydrants are provided along the network at least every 120m to protect T2 and landside buildings.
2. Arrangement of outdoor fire protection pressurized network. The indoor and outdoor fire protection systems of T2 share the pressurized network that meets the requirements for water flow for fire protection at the airport. Consisting of DN300 pipes, the pressurized network is laid around T2 and connected to the indoor fire hydrants and automatic sprinkler system.

7.4.5.3 Design Highlights of Outdoor Water Supply System
1. Highly assured domestic water supply. Dual water sources, namely water supply stations on the south and north of the airport, highly assures the supply of domestic water.
2. The indoor and outdoor fire protection systems shares the pressurized network, which streamlines the fire control system and eliminates the need for outdoor pressurized pipelines for fire protection.
3. The fire protection system of T2 is connected with that of Terminal 1 to guarantee the water supply for fire protection.
4. The outdoor fire protection pressurized system supplies water for outdoor fire protection, reducing the amount of water to be transferred by fire engines.

7.4.6 Municipal Drainage System

7.4.6.1 General
The municipal drainage system includes outdoor domestic sewage and wastewater pipes and rainwater network. The drainage system completely separates rainwater, sewage and wastewater from each other. The feces are treated in the septic tanks and F&B wastewater in grease traps before being discharged into the municipal sewage network.

7.4.6.2 Outdoor Sewage System
1. Drainage network design. As the existing drainage conditions do not permit the laying of gravity sewage pipe in Phase II expansion project, sewage pumping stations are provided under the viaduct on the southwest of T2. The sewage from T2 is collected by the new pumping station and transferred to No. 1 Pumping Station, to No. 2 Pumping Station near the sewage treatment plant, then to the sewage treatment plant for treatment.
2. Arrangement and laying of drainage network. In Phase II, the domestic wastewater from the building and sewage treated in the septic tanks are collected and discharged to the outdoor sewage network. The landside sewage pipelines are laid along the roads around the building, and the airside ones along the service curbsides outside the piers.

7.4.6.3 Outdoor Rainwater System
1. Design parameters. Considering the regional climate and the importance of the project, the design adopts the following parameters: a recurrence interval of 10 years for rainstorms,

that of 50 to 100 years for floods, and runoff coefficient of 0.85.

2. Projection of peak rainwater runoff. According to the analysis of site and building layout, the landside peak rainwater runoff is about 14.87m^3/s and the airside one about 12.9 m^3/s.

3. Outdoor rainwater system. The existing drainage conditions do not permit discharge interfaces in the south and north of Phase II of the expansion project, and stormwater channel tanks on both the east and west to the airside apron become the only option. Based on the communication with the Civil Aviation Administration, the rainwater from the rooftop of T2 on the air side is directly discharged into the stormwater channels near the airside service curbsides. The rainwater from the rooftop of T2 on the land side, that of GTC, and the landside site is collected and discharged to the No. 1 Rainwater Pumping Station in the southwest corner and then pumped to the stormwater channel tanks in the airside aprons.

4. Design Highlights

(1) Rainwater, sewage and wastewater are completely separated to meet the requirements for environmental protection.

(2) A large number of new materials and new processes are utilized to improve the quality of the project and extend the service life of the sewage network.

(3) Large-scale rainwater storage facilities are used to improve the capacity to cope with extreme weather.

7.4.7 Application of Sponge City Concept in T2

7.4.7.1 General
1. Sponge City
The sponge city design provides full protection for natural water bodies and vegetation, minimizes the damage caused by the construction and development to the ecology and environment, effectively reduces the total runoff volume and peak rainfall flow, and eases the pressure on drainage.

2. Water Environment at the Airport

(1) Analysis of vulnerability to waterlogging. The site is a reserved vacant area with the runoff coefficient being less than 0.2. Following the airport development and expansion, the site will be turned into municipal roads, runways, T2 and etc. Due to drastic change with the hardened pavement, the coefficient rises from 0.2 to 0.8, which results in increased total runoff volume and peak rainfall flow and shortened time for rainwater convergence, hence great vulnerability to waterlogging during heavy rains.

(2) Analysis of runoff pollution. As result of airport expansion, a large number of non-point pollution sources such as building clusters and roads will emerge, inevitably leading to a certain degree of runoff pollution.

(3) Analysis of factors limiting runoff. Limited by the drainage conditions, there is no rainwater pipeline in the site of Phase II expansion project. As a result, the rainwater in the landside of T2 has to be pumped to the external river by the No.1 Rainwater Pumping Station. Constrained by the double-hole box culverts, the drainage capacity of the pumping station can only be improved to 12.8m^3/s and the forebay volume of the pumping station cannot be increased. After T2 is expanded, the drainage capacity of the No. 1 Rainwater Pumping Station cannot meet the requirements for drainage in the site;

(4) Analysis of countermeasures. In view of the limited discharge volume and high demand for drainage, flood control, and water storage at the airport, rainwater control and utilization facilities should be used to regulate the peak and total runoff.

7.4.7.2 Sponge City Planning and Design Requirements
1. Overall objects. The concept of sponge city is applied in the airport development to strengthen the capacity for flood control and disaster mitigation, improve the ecology and alleviate the pressure on urban water resources. Based on the goals of drainage, flood control and the development of a "sponge airport" and the characteristics of the water system of the Airport, it is determined that the overall objectives of the project are total runoff control, runoff pollution control, drainage and flood control and protection of water environment.

2. Design strategies. Based on such factors as rainfall and soil, existing issues concerning the water environment, water resources, water ecology, water safety, etc., and requirements for construction, specific strategies implemented are as follows: (1) Processes. Priority is given to bioretention facilities and rainwater gardens that contribute to improving the water quality and controlling runoff. Permeable pavements are applied to replenish groundwater with rainwater on the site. The whole processes focus on the control of the total runoff, and connect with the drainage network and storage tanks to ensure the capacity for drainage and flood control. (2) Control of non-point source pollution. The control of water quality is treated as a critical part of non-point source pollution control, stressing on such indicators of TSS, total nitrogen and total phosphorus. Meanwhile, pollution control is combined with other measures including pollutant interception and dredging to address the issues with water environment governance. (3) Coordination of landscape and environment. Outdoor ecological parking spaces are provided. Vines, as part of the landscaping, are used to improve both the parking area and the capacity of plants for water and soil conservation.

7.4.7.3 Essentials of Sponge City Design
1. Selection of appropriate technologies. Rainwater storage and regulation facilities are used to extend the time for rainwater convergence and reduce the peak runoff; purification ones to mitigate the initial runoff pollution; storage facilities and rainwater transport facilities including grass ditches to utilize rainwater resources.

2. Establishment of the sponge city system.

Stormwater Management Model (SWMM) is used to simulate the site drainage. The dynamics of the runoff at the outlets is calculated by simulating the total runoff and pipeline flow in the site, and based on the calculation results, the effect of low-impact development (LID) facilities in the project is analyzed.
3. Low-impact development strategy. Based on the characteristics of the airport, the LID facilities selected for this project mainly include sunken green spaces, permeable paved roads, rainwater gardens, and planting beds.

7.4.8 Municipal Electrical System

7.4.8.1 Lighting System
1. Design principles of lighting system. (1) The road lighting design mainly focuses on the functions and is part of the landscaping. The road lighting shall be of advanced quality in China and cater to the development requirements and tastes in modern cities. (2) In addition to meeting the requirements for brightness and uniformity of illumination, the road lighting also provide pleasant visibility for drivers. (3) Advocate green lighting and select highly efficient light sources and lamps (0.7 high-pressure sodium lamp / 0.9 LED lamp). The level of protection is not lower than IP65. The shapes of lamps and lamp poles should harmonize with the airport environment.
2. Standards for lighting design. As the project consists of roads, areas for infrastructure and plazas, regional lighting is adopted.
3. Road lighting design. Lights on ground roads and viaducts: (1) As the project includes ground roads, bridges and parking lots, high-mast lights that are coordinated with the surrounding buildings and environment will be used to avoid excessive poles of varied heights. 5X1000W high-pressure sodium lamps (initial lumen of 120000Lm) are installed at the height of 30 meters. (2) Lights on the ground roads under viaducts: 60W LED lamps are attached to the bridge bottom between two piers every 5 meters.

7.4.8.2 Traffic Monitoring and Parking System
1. Traffic monitoring. Monitoring points: The traffic monitoring includes four parts: parking video surveillance, public security and road video surveillance, access control video surveillance, and capture of traffic offenses.
2. Parking management system. (1) Parking areas include parking building, outdoor parking lot, VIP parking lot, bus parking lot, and taxi pick-up area. (2) System features of the parking building and outdoor parking system: access control, automatic license plate recognition at entrances and exits, secondary recognition in the site, indoor parking guidance, reverse car search, vehicle alarm, central charging, self-service charging, manual charging. (3) VIP parking system features: access control and automatic license plate recognition. (4) Bus parking system feature: access control. (5) Taxi pick-up area feature: access control.

7.4.8.3 Municipal Power System
1. The water storage and regulation pumps are of Class II electrical loads. To ensure the reliability of power supply, two low-voltage cabinets of the substation are used to supply power of 380V (one of them is for standby power supply). The pump station is switched on and off with automatic PLCs and grillage machines. Main equipment loads include submersible pumps, sludge pumps, grillage machines, hoists, and backwashing equipment, totaling about 158kw. The calculated load is 108.8kw. Control modes include manual, remote, and automatic control.
2. Technical application. The automatic control system is designed with an open system that is combined with equipment for monitoring. Information flows throughout the production process, enabling easy and reliable exchange of the process control data and information between the PLCs and external facilities. Signal monitoring: The water and sludge gauges in the rainwater storage and regulation tanks transmit all test parameters and equipment operation status to the central control room in real time.

工程大事记

2006.08	签订设计合同
2006~2012	概念方案设计
2008.08	取得国家发改委关于广州白云国际机场扩建工程项目建议书的批复
2012.07	取得国家发改委关于广州白云国际机场扩建工程可行性研究报告的批复
2012.10.25	广东省建筑设计研究院完成初步设计
2012.11.19	初步设计审查会
2012.12.30	广东省建筑设计研究院完成基础施工图设计
2013.04	取得民航局初步设计及概算批复
2013.04.28	二号航站楼项目奠基仪式
2013.06.30	广东省建筑设计研究院完成航站楼市政道路工程施工图设计
2013.08.30	广东省建筑设计研究院完成交通中心施工图设计
2013.9.30	广东省建筑设计研究院完成航站楼土建施工图设计
2013.12.30	广东省建筑设计研究院完成航站楼机电施工图设计
2013.12.30	广东省建筑设计研究院完成航站楼市政桥梁、隧道工程施工图设计
2014.02	航站楼主体结构开工
2014.04	北进场隧道开工
2014.04.30	广东省建筑设计研究院完成交通中心机电施工图设计
2014.06	交通中心、城轨、地铁工程开工
2014.07.20	广东省建筑设计研究院完成交通中心土建施工图设计
2014.11.15	广东省建筑设计研究院完成航站楼幕墙施工图设计
2015.03	航站楼跨线桥工程开工
2015.03.10	广东省建筑设计研究院完成航站楼屋面施工图设计
2015.04.10	广东省建筑设计研究院完成航站楼张拉膜雨篷施工图设计
2015.07	航站楼钢结构工程开工
2015.09	航站楼幕墙工程开工
2015.10	航站楼屋面工程开工
2015.11	航站楼机电工程开工
2015.12.10	广东省建筑设计研究院完成航站楼及交通中心装修施工图设计
2016.01	行李处理系统采购项目进场
2016.01	完成土建工程施工
2016.01	完成桩基础工程施工
2016.03	航站楼及配套设施地面石材项目进场施工
2016.03	应急电源装置（EPS）项目进场
2016.03	自动感应门项目进场
2016.03	冷水机组及冷源群控系统项目进场
2016.03	航站楼安防工程进场
2016.05	离港控制等系统工程和机场运行信息集成平台及应用系统工程进场
2016.07	室内装修工程开工
2016.08	航站楼主体结构已基本完成，钢结构、玻璃幕墙、屋面已完成；机电安装完成过半，弱电工程已按计划开工
2016.10.30	北进场隧道启用
2017.06.15	白云机场交通中心及停车楼项目主体结构提前封顶
2017.08.01	白云机场扩建工程航站区站坪及配套设施通过竣工验收
2017.08.13	随着白云机场扩建工程二号航站楼出港高架桥最后一跨第五联桥面混凝土浇筑完成，宣告了白云机场扩建工程出港高架桥顺利合拢
2017.08.17-18	白云机场扩建工程航站区站坪及配套设施通过行业验收
2017.10.12	白云机场二号航站楼停机坪正式启用
2017.11.15	白云机场二号航站楼进入弱电系统、机电系统现场联合调试阶段
2018.02.06~07	白云机场扩建工程二号航站楼及配套设施工程通过工程竣工验收
2018.02.9-10	白云机场扩建工程二号航站楼及配套设施工程通过民航中南管理局行业验收
2018.04.26	白云机场二号航站楼、交通中心、站坪及配套设施正式启用
2018.05.19	南航全面转场进驻二号航站楼

MILESTONES

Aug. 2006: Conclusion of design contract
2006-2012: Conceptual design
Aug. 2008: NDRC approves the Project Proposal
Jul. 2012: NDRC approves the Project Feasibility Study Report
Oct. 25, 2012: GDAD completes preliminary design
Nov. 19-23, 2012: Preliminary design review meeting
Dec. 30, 2012: GDAD completes foundation CD design
Apr. 2013: CAAC approves the preliminary design and budgetary estimate
Apr. 2013: Foundation laying ceremony of T2
Jun. 30, 2013: GDAD completes municipal road CD design of T2
Aug. 30, 2013: GDAD completes CD design of GTC
Sep. 30, 2013: GDAD completes civil engineering CD design of T2
Dec. 30, 2013: GDAD completes MEP CD design of T2
Dec. 30, 2013: GDAD completes municipal bridge and tunnel CD design of T2
Feb. 2014: Commencement of main structure works of T2
Apr. 2014: Commencement of North Approach Tunnel works
Apr. 30, 2014: GDAD completes MEP CD design of GTC
Jun. 2014: Commencement of GTC, intercity railway and metro works
Jul. 20, 2014: GDAD completes civil engineering CD design of GTC
Nov. 15, 2014: GDAD completes curtain wall CD design of T2
Mar. 2015: Commencement of flyover works of T2
Mar. 10, 2015: GDAD completes roof CD design of T2
Apr. 10, 2015: GDAD completes tensile membrane canopy CD design of T2
Jul. 2015: Commencement of steel structure works of T2
Sep. 2015: Commencement of curtain wall works of T2
Oct. 2015: Commencement of roof works of T2
Nov. 2015: Commencement of MEP works of T2
Dec. 10, 2015 GDAD completes finish CD design of T2 and GTC
Jan. 2016: Baggage handling system mobilizes into the site
Jan. 2016: Completion of civil construction
Jan. 2016: Completion of pile foundation construction
Mar. 2016: Ground stone materials for T2 and supporting facilities mobilize into the site
Mar. 2016: Emergency power supply (EPS) units mobilize into the site
Mar. 2016: Automatic doors mobilizes into the site
Mar. 2016: Water chilling unit and cold source group control system mobilizes into the site
Mar. 2016: Security system mobilizes into the site
May 2016: Departure control and other systems as well as airport operation information integration platform and application system mobilize into the site
Jul. 2016: Commencement of interior finish works
Aug. 2016: Fundamental completion of the main structure and completion of steel structure, glass curtain wall and roof works of T2; half completion of MEP works and commencement of ELV works
Oct. 30, 2016 : The north approach tunnel is put into use
Jun. 15, 2017: Ahead-of-schedule completion of the main structure of GTC and parking building
Aug. 1, 2017: Completion acceptance of terminal area apron and supporting facilities
Aug. 13, 2017: Completion of final concrete pouring on the fifth deck of the last span of the departure viaduct of T2
Aug. 17-18, 2017: Industry acceptance of the terminal area apron and supporting facilities
Oct. 12, 2017: Official opening of the apron of T2
Nov. 15, 2017: On-site joint commissioning of ELV and MEP systems of T2
Feb. 6-7, 2018: Completion acceptance of T2 and supporting facilities
Feb. 9-10, 2018: Industry acceptance of T2 and supporting facilities by CAAC Central and Southern Regional Administration
Apr. 26, 2018: Official opening of T2, GTC, apron and supporting facilities
May 19, 2018: Mobilization of China Southern Airlines into T2

广东省建筑设计研究院项目设计团队

项目主持人：	何锦超（2006~2012）、王洪（2012~2014）、赏锦国（2015~2017）、陈朝阳（2012~2018）、陈雄（2017~2018）
项目总负责：	陈雄（2006~2018）、潘勇（2013~2018）周昶（2010~2017）、郭胜（2006~2009）
建筑专业负责人：	赖文辉、郭其轶、易田、钟伟华
建筑分项负责人：	邓章豪、杨坤、吴冠宇
审定人：	陈雄
审核人：	黄志东、李琦真
建筑专业全体设计人：	陈雄、潘勇、周昶、赖文辉、郭其轶、易田、钟伟华、邓章豪、杨坤、吴冠宇、温云养、罗菲、许尧强、戴志辉、倪俍、董轩、金少雄、黎智立、黎运武
结构专业负责人：	陈星、区彤、李桢章、李恺平
结构分项负责人：	谭坚、傅剑波、谭和
审定人：	陈星
审核人：	李桢章
结构专业全体设计人：	陈星、区彤、李桢章、李恺平、谭坚、傅剑波、谭和、劳智源、戴鹏森、张连飞、王艳霞、刘雪兵、张艳辉、刘星兰、张增球、林家豪、张鸿雁、陈前、罗益群、温惠祺、潘浩彦、杨飞
电气专业负责人：	陈建飚、钟世权
电气分项负责人：	黄日带、李村晓、周小蔚、何军、赵骥
审定人：	陈建飚
审核人：	庄孙毅、黄宇清
电气专业全体设计人：	陈建飚、钟世权、何军、黄日带、李村晓、周小蔚、赵骥、崔明丽、张文武、汤志标、王飞、周一锋、肖幸怀、卞嘉铭、徐志钢、简永杰、张敏、钟奇宏、曾祥、周宇健、郭永杰、曾建敏、黄友树、熊洁、苏俊勇、方智
空调专业负责人：	廖坚卫、陈小辉
空调分项负责人：	何花、赖文彬、郭林文
审定人：	廖坚卫
审核人：	陈小辉
空调专业全体设计人：	廖坚卫、陈小辉、赖文彬、何花、郭林文、许杰、方标、邹永胜、莫煜均、谭思为、钟铿、李进、屈永强、何妞、廖捷
给排水专业负责人：	符培勇、梁景晖、黎洁
给排水分项负责人：	赖振贵、普大华、彭康
审定人：	符培勇
审核人：	叶志良、罗谨
给排水专业全体设计人：	符培勇、梁景晖、黎洁、赖振贵、普大华、彭康、林兆铭、钟可华、李东海、陈文杰、何慧婷、张伟、王军慧
市政专业负责人：	陈伟
市政分项负责人：	黎洁、许海峰、陈东哲、刘明、纪鹏、胡智敏、饶欣频、邬龙刚、周惠明
审定人：	曹旭华、陈伟、符培勇、李来埔、廖坚卫
审核人：	罗美辉、陈伟、罗谨、刘明辉、沈洪
市政专业全体设计人：	陈伟、黎洁、许海峰、刘明、纪鹏、胡智敏、饶欣频、邬龙刚、周惠明、金学锋、吴敏、张正军、顾庆福、丁伟亮、陈海斌、尹华、崔剑奇、安丽勇、曹敏辉、张强、黄龙田、王华松、刘光明、彭康、何慧婷、张伟、黄洁莹、冯刚、李青、孔庆存
景观专业负责人：	彭国兴、林娜、许海峰、刘亚超、陈晓齐
景观分项负责人：	陈颖
审定人：	郭奕辉、李来埔、黄维让、甄宏志
审核人：	陈颖、彭国兴、刘明辉、焦瑞虎、陈位洪
景观专业全体设计人：	彭国兴、林娜、许海峰、刘亚超、陈晓齐、郭奕辉、李来埔、黄维让、甄宏志、陈颖、刘明辉、焦瑞虎、陈位洪、石超、张伊、黄洁莹、周兴、陈晓齐
预算专业负责人：	许春燕、许爱斌
预算专业分项负责人：	许春燕、许爱斌
审定人：	刘毅红、许春燕
审核人：	潘富成、许春燕、余业辉
预算专业全体设计人：	许春燕、许爱斌、陈少华、潘富成、马翠娟、邓乐、宋琴、严国平、潘雯莉、连健、谢宏南、张文坚、刘永光、沈师师、袁颖意、戴业琪、方华颖
项目秘书：	宋翠芬

DESIGN TEAM OF GDAD

Project Directors: He Jinchao (2006-2012), Wang Hong (2012-2014), Shang Jinguo (2015-2017), Chen Chaoyang (2012-2018), Chen Xiong (2017-2018)
Project Principals: Chen Xiong (2006-2018), Pan Yong (2013-2018), Zhou Chang (2010-2017), Guo Sheng (2006-2009)
Architectural Principals: Lai Wenhui, Guo Qiyi, Yi Tian, Zhong Weihua
Architectural Supervisors: Deng Zhanghao, Yang Kun, Wu Guanyu
Approver: Chen Xiong
Reviewer: Huang Zhidong, Li Qizhen
Architectural Team: Chen Xiong, Pan Yong, Zhou Chang, Lai Wenhui, Guo Qiyi, Yi Tian, Zhong Weihua, Deng Zhanghao, Yang Kun, Wu Guanyu, Wen Yunyang, Luo Fei, Xu Yaoqiang, Dai Zhihui, Ni Liang, Dong Xuan, Jin Shaoxiong, Li Zhili, Li Yunwu

Structural Principals: Chen Xing, Ou Tong, Li Zhenzhang, Li Kaiping
Structural Supervisors: Tan Jian, Fu Jianbo, Tan He
Approver: Chen Xing
Reviewer: Li Zhenzhang
Structural Team: Chen Xing, Ou Tong, Li Zhenzhang, Li Kaiping, Tan Jian, Fu Jianbo, Tan He, Lao Zhiyuan, Dai Pengsen, Zhang Lianfei, Wang Yanxia, Liu Xuebing, Zhang Yanhui, Liu Xinglan, Zhang Zengqiu, Lin Jiahao, Zhang Hongyan, Chen Qian, Luo Yiqun, Wen Huiqi, Pan Haoyan, Yang Fei

Electrical Principals: Chen Jianbiao, Zhong Shiquan
Electrical Supervisors: Huang Ridai, Li Cunxiao, Zhou Xiaowei, He Jun, Zhao Ji
Approver: Chen Jianbiao
Reviewers: Zhuang Sunyi, Huang Yuqing
Electrical Team: Chen Jianbiao, Zhong Shiquan, He Jun, Huang Ridai, Li Cunxiao, Zhou Xiaowei, Zhao Ji, Cui Mingli, Zhang Wenwu, Tang Zhibiao, Wang Fei, Zhou Yifeng, Xiao Xinghuai, Bian Jiaming, Xu Zhigang, Jian Yongjie, Zhang Min, Zhong Qihong, Zeng Xiang, Zhou Yujian, Guo Yongjie, Zeng Jianmin, Huang Youshu, Xiong Jie, Su Junyong, Fang Zhi

AC Principals: Liao Jianwei, Chen Xiaohui
AC Supervisors: He Hua, Lai Wenbin, Guo Linwen
Approver: Liao Jianwei
Reviewer: Chen Xiaohui
AC Team: Liao Jianwei, Chen Xiaohui, Lai Wenbin, He Hua, Guo Linwen, Xu Jie, Fang Biao, Zou Yongsheng, Mo Yujun, Tan Siwei, Zhong Keng, Li Jin, Qu Yongqiang, He Niu, Liao Jie

Plumbing Principals: Fu Peiyong, Liang Jinghui, Li Jie
Plumbing Supervisors: Lai Zhengui, Pu Dahua, Peng Kang
Approver: Fu Peiyong
Reviewers: Ye Zhiliang, Luo Jin
Plumbing Team: Fu Peiyong, Liang Jinghui, Li Jie, Lai Zhengui, Pu Dahua, Peng Kang, Lin Zhaoming, Zhong Kehua, Li Donghai, Chen Wenjie, He Huiting, Zhang Wei, Wang Junhui

Municipal Principal: Chen Wei
Municipal Supervisors: Li Jie, Xu Haifeng, Chen Dongzhe, Liu Ming, Ji Peng, Hu Zhimin, Rao Xinpin, Wu Longgang, Zhou Huiming
Approvers: Cao Xuhua, Chen Wei, Fu Peiyong, Li Laipu, Liao Jianwei
Reviewers: Luo Meihui, Chen Wei, Luo Jin, Liu Minghui, Shen Hong
Municipal Team: Chen Wei, Li Jie, Xu Haifeng, Liu Ming, Ji Peng, Hu Zhimin, Rao Xinpin, Wu Longgang, Zhou Huiming, Jin Xuefeng, Wu Min, Zhang Zhengjun, Gu Qingfu, Ding Weiliang, Chen Haibin, Yin Hua, Cui Jianqi, An Liyong, Cao Minhui, Zhang Qiang, Huang Longtian, Wang Huasong, Liu Guangmin, Peng Kang, He Huiting, Zhang Wei, Huang Jieying, Feng Gang, Li Qing, Kong Qingcun

Landscape Principals: Peng Guoxing, Lin Na, Xu Haifeng, Liu Yachao, Chen Xiaoqi
Landscape Supervisor: Chen Ying
Approvers: Guo Yihui, Li Laipu, Huang Weirang, Zhen Hongzhi
Reviewers: Chen Ying, Peng Guoxing, Liu Minghui, Jiao Ruihu, Chen Weihong
Landscape Team: Peng Guoxing, Lin Na, Xu Haifeng, Liu Yachao, Chen Xiaoqi, Guo Yihui, Li Laipu, Huang Weirang, Zhen Hongzhi, Chen Ying, Liu Minghui, Jiao Ruihu, Chen Weihong, Shi Chao, Zhang Yi, Huang Jie ying, Zhou Xing, Chen Xiaoqi

Budget Principals: Xu Chunyan, Xu Aibin
Budget Supervisors: Xu Chunyan, Xu Aibin
Approvers: Liu Yihong, Xu Chunyan
Reviewers: Pan Fucheng, Xu Chunyan, Yu Yehui
Budget Team: Xu Chunyan, Xu Aibin, Chen Shaohua, Pan Fucheng, Ma Cuijuan, Deng Le, Song Qin, Yan Guoping, Pan Wenli, Lian Jian, Xie Hongnan, Zhang Wenjian, Liu Yongguang, Shen Shishi, Yuan Yingyi, Dai Yeqi, Fang Huaying
Project Secretary : Song Cuifen

广东省建筑设计研究院项目设计团队
DESIGN TEAM OF GDAD

陈雄　Chen Xiong

陈雄先生1983年华南工学院建筑学系工学学士毕业。1986年华南工学院建筑设计研究院建筑理论与设计硕士研究生毕业，同年进入广东省建筑设计研究院。1999年起担任广东省建筑设计研究院副总建筑师，2004年筹建机场设计所并担任所长10年，2015年起担任广东省建筑设计研究院副院长兼总建筑师。2016年被评为全国工程勘察设计大师。是教授级高级建筑师、中国一级注册建筑师。获得"当代中国百名建筑师"、"全国建设系统先进工作者"、"广东省建设系统先进工作者"、"广东省五一劳动奖章"等荣誉称号。曾荣获全国优秀工程设计金奖。2006年至2018年全过程担任广州白云国际机场二号航站楼及配套设施项目总负责人，全面主持设计工作，组织跨部门多专业的设计团队，完成从方案设计到施工图设计及现场服务。对外处理与业主等参建各方的协调，对内就工程设计及建设过程中的重大技术方案、技术难题制定或审查解决方案，确保工程高质量完成。

Graduated with a Bachelor's Degree in Engineering in 1983 and a Master's Degree in Architectural Theory and Design in 1986 from the South China Institute of Technology (now SCUT). Worked with GDAD since 1986 and served as the associate chief architect since 1999. Set up the ADG in 2004 and directed the office for 10 years. Served as the vice president and chief architect of GDAD since 2015 and elected as a National Engineering Survey and Design Master in 2016. He is a professor-level senior architect and a first-class registered architect in China. Honors received include Top 100 Contemporary Architects in China, National/Provincial Advanced Worker of Building Industry, and the Provincial May 1st Labor Medal. Winner of National Project of Design Excellence (Gold Prize). As project director of T2 and its supporting facilities of the Baiyun Int'l Airport during 2006-2018, led the cross-departmental multidisciplinary design team and directed all design works throughout SD, CD and CA. Coordinated with client and other project teams and developed/reviewed technical solutions for key issues at design and construction stages to ensure the high-quality realization of the project.

潘勇　Pan Yong

潘勇先生1994年华南理工大学建筑学系建筑学学士毕业。1997年进入广东省建筑设计研究院。2004年起担任省院机场设计所副总建筑师。2009年起担任省院机场设计所总建筑师，2016年起担任广东省建筑设计研究院副总建筑师。是教授级高级建筑师、中国一级注册建筑师、AAA2014亚洲建筑师协会奖荣誉奖获得者、第八届中国建筑学会青年建筑师、广东省十佳中青年建筑师。2013年至2018年担任广州白云国际机场二号航站楼及配套设施项目主创建筑师及总负责人，主创整体造型及空间形态创意设计，负责建筑各系统整合设计及项目组织协调工作。

Graduated with a Bachelor's Degree in Architecture in 1994 from SCUT and worked with GDAD since 1997. Served as the associate chief architect and the chief architect of the ADG since 2004 and 2009 respectively, and the associate chief architect of GDAD since 2016. He is a professor-level senior architect, a first-class registered architect in China, winner of 2014 ARCASIA Awards for Architecture (Honorable Mention), the 8th ASC Young Architect Awards, and the Top 10 Young and Middle-aged Architect Awards of Guangdong. As chief creative architect and project director of T2 and its supporting facilities of Baiyun Int'l Airport from 2013 to 2018, directed creative design of building and spatial forms and coordinated integrated designs of building systems and project organization.

周昶　Zhou Chang

周昶先生1994年华南理工大学建筑学系建筑学学士毕业。1997年进入广东省建筑设计研究院。2004年起担任广东省建筑设计研究院机场设计所副所长。是建筑学高级工程师。2010年至2017年担任广州白云国际机场二号航站楼及配套设施项目总负责人，主创平面布局与旅客流程设计，负责土建及围护系统技术设计，配合项目前期总体筹划、规划布局与构型设计，协调与业主及各参建单位的对接工作，推进整个设计流程与项目管理与建设工作的顺利对接。

Graduated with a Bachelor's Degree in Architecture in 1994 from SCUT and worked with GDAD since 1997. Served as deputy director of ADG since 2004. He is a senior engineer in architecture. As project director of T2 and its supporting facilities of Baiyun Int'l Airport from 2010 to 2017, directed planar layout and passenger flow design and technical design of civil works and envelops, supported overall planning, terminal planning and layout, and coordinated the contacts with client and other project teams to promote the smooth connection of project design, management and construction.

赖文辉　Lai Wenhui

赖文辉先生2006年广州大学建筑与城市规划学院建筑学学士毕业，同年进入广东省建筑设计研究院。2016年起担任广东省建筑设计研究院机场设计所建筑室主任。是建筑学高级工程师。2012年至2018年担任广州白云国际机场扩建工程二号航站楼及配套设施建筑设计专业负责人，负责航站楼主楼建筑及装修设计、旅客流程设计、联检设计、行李系统及APM系统等相关设计。

Graduated from GZHU with a Bachelor's Degree in Architecture and joined GDAD in 2006; served as the Director of Architecture Office of ADG since 2016. A senior architect. An architectural principal of T2 and its supporting facilities of the Baiyun Int'l Airport during 2012-2018 in charge of architecture, finish and passenger flow design of the main building of T2.

郭其轶　Guo Qiyi

郭其轶先生2007年华中科技大学建筑与城市规划学院建筑学硕士研究生毕业，同年进入广东省建筑设计研究院。2016年起担任广东省建筑设计研究院机场设计所建筑室副主任。是建筑学高级工程师、中国一级注册建筑师。2012年至2018年担任广州白云国际机场扩建工程二号航站楼及配套设施建筑设计专业负责人，负责航站楼指廊建筑及装修设计、贵宾室建筑及装修设计、标识系统、座椅、空侧室外工程等相关设计。

Graduated from HUST with a Master's Degree in Architecture and joined GDAD in 2007; serves as the Deputy Director of Architecture Office of ADG since 2016. A senior architect and a first-class registered architect in China. An architectural principal of T2 and its supporting facilities of the Baiyun Int'l Airport during 2012-2018 in charge of architecture and finish design of piers and VIP rooms in T2.

易田　Yi Tian

易田先生2007年华南理工大学建筑学系建筑学学士毕业，2009年进入广东省建筑设计研究院。2016年起担任广东省建筑设计研究院机场设计所/ADG建筑创作工作室主任建筑师。是建筑学高级工程师、中国一级注册建筑师。2014年至2018年担任广州白云国际机场扩建工程二号航站楼及配套设施建筑专业负责人，负责航站楼建筑方案及商业区装修施工图设计、办票岛、柜台以及栏杆隔断系统等相关设计。

Graduated from SCUT with a Bachelor's Degree in Architecture in 2007; joined GDAD in 2009; served as principal architect in ADG / ADG Architectural Design Institute since 2016. A senior architecture and a first-class registered architect in China. An architectural principal of T2 and its supporting facilities of the Baiyun Int'l Airport during 2014-2018 in charge of architectural design and commercial area finish CD design of T2.

钟伟华　Zhong Weihua

钟伟华女士2002年广州大学建筑学工学学士毕业，2010年进入广东省建筑设计研究院，2017年起担任广东省建筑设计研究院机场设计所主任建筑师。是建筑学高级工程师、中国一级注册建筑师。2012年至2018年担任广州白云国际机场扩建工程二号航站楼及配套设施建筑专业负责人，负责交通中心及停车楼建筑及装修设计、航站楼柜台、机房装修以及陆侧室外工程等相关设计。

Graduated from GZHU with a Bachelor's Degree in Architecture in 2002; joined GDAD in 2010; served as principal architect in ADG. A senior architect and a first-class registered architect in China. An architectural principal of T2 and its supporting facilities of the Baiyun Int'l Airport during 2012-2018 in charge of architectural and finish design of GTC and parking building and counter design of T2.

广东省建筑设计研究院项目设计团队
DESIGN TEAM OF GDAD

陈星　Chen Xing

陈星先生1982年华南工学院建筑结构专业本科毕业，同年进入广东省建筑设计研究院，现任广东省工程勘察设计行业协会会长、广东省建筑设计研究院结构专业顾问总工程师。是教授级高级工程师、国家一级注册结构工程师、中港互认结构工程师、国务院政府特殊津贴专家、广东省工程勘察设计大师、华南理工大学兼职教授、全国超限高层建筑工程抗震设防审查专家委员会委员、中国勘察设计协会副理事长、中国建筑学会建筑结构分会副理事长、广东省土木建筑学会副理事长等。2012年至2018年担任广州白云国际机场二号航站楼及配套设施结构专业负责人。

Graduated from SCUT with a Bachelor's Degree in Architecture and joined GDAD in 1982. The president of Guangdong Engineering Survey and Design Industry Association and the chief engineer of the structure discipline of GDAD. A professor-level senior engineer, a first-class registered structural engineer in China, a Sino-Hong Kong Mutual Recognition Structural Engineer, a State Council Special Allowance Expert, a Guangdong Provincial Engineering Survey and Design Master, an adjunct professor at SCUT. A structural design principal of T2 and its supporting facilities of the Baiyun Int'l Airport during 2012-2018.

区彤　Ou Tong

区彤先生1995年华南理工大学工业与民用建筑专业本科毕业，2008年进入广东省建筑设计研究院。2018年起担任广东省建筑设计研究院机场设计所结构总工程师，2019年起担任广东省建筑设计研究院结构专业副总工程师。是高级工程师、中国一级注册结构工程师、广东省建筑金属围护系统工程技术研究中心主任。2012年至2018年担任广州白云国际机场二号航站楼及配套设施结构专业负责人，负责结构专业的各项技术和计划具体执行，并主持"白云国际机场二号航站楼建造关键技术"相关科研工作。

Graduated from SCUT with a Bachelor's Degree in Industrial and Civil Architecture in 1995; joined GDAD in 2008; chief structural engineer of ADG since 2018 Associate chief structural engineer of GDAD since 2019. A senior engineer, a first-class registered structural engineer in China, and the director of Guangdong Engineering Technology Research Center for Metal Building Envelop System. A structural design principal of T2 and its supporting facilities of the Baiyun Int'l Airport during 2012-2018; conducted scientific research on Key Construction Technologies for T2 of the Baiyun Int'l Airport.

李桢章　Li Zhenzhang

李桢章先生是华南理工大学建筑结构工程力学学士，1975年进入广东省建筑设计院从事建筑结构设计工作。2004年起担任广东省建筑设计研究院副总工程师，现任广东省建筑设计研究院审图中心顾问总工程师。是教授级高级工程师、一级注册结构工程师、国务院特殊津贴专家、中国建筑学会资深会员、广东省超限工程审查专家。2012年至2018年担任广州白云国际机场二号航站楼及配套设施结构专业负责人。

Graduated from SCUT with a Bachelor's Degree in Engineering in the major of Building Structure Engineering Mechanics; joined GDAD in 1975; served as deputy chief engineer of GDAD since 2004 and is the chief consulting engineer of the Drawing Review Center of GDAD. A professor-level senior engineer, a first-class registered structural engineer in China, and a State Council Special Allowance Expert. A structural design principal of T2 and its supporting facilities of the Baiyun Int'l Airport during 2012-2018.

李恺平　Li Kaiping

李恺平先生1989年南京建筑工程学院工业与民用建筑专业工学学士本科毕业，1993年同济大学结构工程专业硕士研究生毕业。1999年进入广东省建筑设计研究院。2016年任省院钢结构设计研究中心副主任兼院结构专业副总工程师。是教授级高级工程师、中国一级注册结构工程师。2019年被评为广东钢结构事业二十五年杰出个人。2012年至2018年担任广州白云国际机场二号航站楼及配套设施结构专业负责人。

Graduated from Nanjing Institute of Architectural Engineering with a Bachelor's Degree in Engineering in the major of Industrial and Civil Architecture in 1989 and from TJU with a Master's Degree in structural engineering in in 1993; joined GDAD in 1999; served as deputy director of the Steel Structure Design Research Center and deputy chief engineer of the structure discipline of GDAD in 2016. A professor-level senior engineer and a first-class registered structural engineer in China. Named an outstanding individual of 25-year Steel Structure Undertaking of Guangdong in 2019. A structural design principal of T2 and its supporting facilities of the Baiyun Int'l Airport during 2012-2018.

陈建飚　Chen Jianbiao

陈建飚先生1984年重庆建筑工程学院工业电气自动化专业毕业。同年进入广东省建筑设计研究院。2004年起担任省院电气专业副总工程师，2013年起担任省院电气专业总工程师，2017年起担任省院副院长兼电气专业总工程师。是教授级高级工程师、注册电气工程师（供配电），中国勘察设计协会建筑电气工程设计分会副会长、中国建筑学会建筑电气分会副理事长等。2012年至2018年担任广州白云国际机场二号航站楼及配套设施电气专业负责人。

Graduated from Chongqing Institute of Civil Engineering and Architecture and joined GDAD in 1984; deputy chief electrical engineer since 2004, chief electrical engineer since 2013, and vice president and chief electrical engineer since 2017 of GDAD. A professor-level senior engineer, a registered electrical (power supply and distribution) engineer, and vice president of Building Electrical Engineering Design Branch of China Engineering & Consulting Association. A structural design principal of T2 and its supporting facilities of the Baiyun Int'l Airport during 2012-2018.

钟世权　Zhong Shiquan

钟世权先生2001年广东工业大学电气技术专业毕业，同年进入广东省建筑设计研究院。现任广东省建筑设计研究院第三机电研究所副所长。是教授级高级工程师、注册电气工程师（供配电）、中国勘察设计协会建筑电气工程设计分会电气专业杰出青年工作组委员、中国勘察设计协会建筑电气工程设计分会华南地区学会理事、广州市城市照明协会副会长等。2012年至2018年担任广州白云国际机场二号航站楼及配套设施电气专业负责人及机电协调人。

Graduated from the major of electrical technology of GDUT and joined GDAD in 2001; the deputy director of the 3rd M&E Research Institute of GDAD. A professor-level senior engineer, a registered electrical (power supply and distribution) engineer, a member of the Electrical Engineering Distinguished Youth Working Group & a director of the South China Society of Building Electrical Engineering Design Branch of China Engineering & Consulting Association. An electrical principal and an MEP coordinator of T2 and its supporting facilities of the Baiyun Int'l Airport during 2012-2018.

廖坚卫　Liao Jianwei

廖坚卫先生1982年重庆建筑工程学院供热与通风专业学士毕业，同年进入广东省建筑设计研究院，从事暖通空调专业设计及技术管理工作至今。2005年起任广东省建筑设计研究院暖通空调副总工程师。是教授级高级工程师、注册设备（暖通空调）工程师。参与《建筑防排烟技术规范》等标准编制。2012年至2018年担任广州白云国际机场二号航站楼及配套设施空调专业负责人，担任暖通空调专业负责及技术审定工作。

Graduated from Chongqing Institute of Civil Engineering and Architecture with a Bachelor's Degree in the major of Heating and Ventilation and joined GDAD for HVAC design and technical management in 1982; deputy chief HVAC engineer since 2005 in GDAD. A professor-level senior engineer and a registered equipment (HVAC) engineer. An AC principal of T2 and its supporting facilities of the Baiyun Int'l Airport during 2012-2018.

陈小辉　Chen Xiaohui

陈小辉女士1986年上海同济大学供热通风与空气调节专业学士毕业，同年进入广东省建筑设计研究院，从事暖通空调专业设计工作和审图工作至今。现任广东省建筑设计研究院机电三所空调专业总工程师。是教授级高级工程师、注册设备（暖通空调）工程师。2012年至2018年担任广州白云国际机场二号航站楼及配套设施空调专业负责人及执行工种负责。

Graduated from TJU with a Bachelor's Degree in the major of HVAC and joined GDAD for HVAC design and drawing review in 1986. The chief AC engineer of the 3rd M&E Research Institute of GDAD. A professor-level senior engineer and a registered equipment (HVAC) engineer. An AC principal and a supervisor of T2 and its supporting facilities of the Baiyun Int'l Airport during 2012-2018.

符培勇　Fu Peiyong

符培勇先生1982年重庆建筑工程学院城建系给水排水专业工学士毕业，同年进入广东省建筑设计研究院，从事给水排水工程设计和技术管理工作。现任广东省建筑设计研究院顾问总工程师。是教授级高级工程师、注册设备（给水排水）工程师、中国建筑学会建筑给水排水技术研究分会名誉理事，中国土木工程学会水工业分会建筑给水排水委员会资深委员等。2012年至2018年担任广州白云国际机场二号航站楼及配套设施给排水专业负责人。

Graduated from Chongqing Institute of Civil Engineering and Architecture with a Bachelor's Degree in Engineering for the major of Water Supply and Drainage and joined GDAD in 1982; currently the chief engineer of GDAD. A professor-level senior engineer, a registered equipment (plumbing) engineer, an honorary director of Building Plumbing Technology Research Branch of the Architectural Society of China. A plumbing principal of T2 and its supporting facilities of the Baiyun Int'l Airport during 2012-2018.

梁景晖　Liang Jinghui

梁景晖先生广州大学建筑工程系给水排水工程专业毕业，1988年进入广东省建筑设计研究院。2013年起担任广东省建筑设计研究院机场设计所给排水专业总工程师。是教授级高级工程师、广东省建筑设计研究院给排水专业技管组成员、广东省综合评标评审专家库专家。2012年至2018年担任广州白云国际机场二号航站楼及配套设施给排水专业负责人。

Graduated from the major of Plumbing Engineering in the department of Architectural Engineering of GZHU and joined GDAD in 1988; chief plumbing engineer of ADG since 2013. A professor-level senior engineer, and a member of GDAD's plumbing technical management team. A plumbing principal of T2 and its supporting facilities of the Baiyun Int'l Airport during 2012-2018.

广东省建筑设计研究院项目设计团队
DESIGN TEAM OF GDAD

陈伟　Chen Wei

陈伟先生2000年上海同济大学交通土建专业毕业，同年进入广东省建筑设计研究院。2009年起担任广东省建筑设计研究院市政一所副总工程师，2015年起担任广东省建筑设计研究院市政道桥副总工程师。是教授级高级工程师、广州市建设科技委市政专业专家。工作期间共参加50余项区域路网、特大桥梁、城市长隧道、城市地下道路系统等不同类型市政工程项目设计及科研。2012年至2018年担任广州白云国际机场二号航站楼及配套设施市政专业负责人。

Graduated from the major of Traffic Civil Engineering of TJU and joined GDAD in 2000; deputy chief engineer of the 1st Municipal Design Institute of GDAD since 2009 and deputy chief engineer for municipal road and bridge design of GDAD since 2015. A professor-level senior engineer and a municipal design expert of Guangzhou Construction Science and Technology Commission. A municipal principal of T2 and its supporting facilities of the Baiyun Int'l Airport during 2012-2018.

黎洁　Li Jie

黎洁先生2000年重庆大学城市建设学院给水排水工程专业毕业，同年进入广东省建筑设计研究院从事给排水设计工作。2015年起担任广东省建筑设计研究院第二机电设计研究所副所长。是教授级高级工程师，注册设备（给水排水）工程师、中国建筑给水排水青年工程师委员会副主任委员。2012年至2018年担任广州白云国际机场二号航站楼及配套设施给排水专业负责人。

Graduated from the major of Plumbing Engineering in the School of Urban Construction of Chongqing University and joined GDAD in 2000; deputy director of the 2nd M&E Design Institute of GDAD since 2015. A professor-level senior engineer, a registered equipment (plumbing) engineer, and a deputy director member of Young Engineers Committee of Building Plumbing Technology Research Branch of ASC. A plumbing principal of T2 and its supporting facilities of the Baiyun Int'l Airport during 2012-2018.

彭国兴　Peng Guoxing

彭国兴先生1993年湘潭建筑学院工民建专业毕业，2008年进入广东省建筑设计研究院。2012起担任广东省建筑设计研究院第一城市与景观设计研究所一室主任，2019年起担任入广东省建筑设计研究院第一城市与景观设计研究所园林专业副总工程师。是高级工程师。2012年至2018年担任广州白云国际机场二号航站楼及配套设施景观设计绿化专业负责人。

Graduated from the Industrial and Civil Construction Engineering major of Xiangtan School of Architecture in 1993. Joined GDAD in 2008; director of the 1st Design Studio of the 1st City and Landscape Design Institute of GDAD since 2012, and deputy chief engineer (landscape architecture) of the 1st City and Landscape Design Institute of GDAD since 2019. A senior engineer. A landscape principal (greening) of T2 and its supporting facilities of the Baiyun Int'l Airport during 2012-2018.

林娜　lin Na

林娜女士2006年毕业于北京林业大学园林专业。2013年进入广东省建筑设计研究院。2015年起担任广东省建筑设计研究院第一城市与景观设计研究所设计室副主任工程师。2015年至2018年担任广州白云国际机场二号航站楼及配套设施景观设计园建专业负责人。

Graduated from the Landscape Architecture major of Beijing Forestry University in 2006. Joined GDAD in 2013; a senior staff engineer of the Design Studio of the 1st City and Landscape Design Institute of GDAD since 2015. A landscape principal (garden architecture) of T2 and its supporting facilities of the Baiyun Int'l Airport during 2015-2018.

许春燕　Xu Chunyan

许春燕女士1993年重庆建筑工程学院建筑管理工程系工学学士毕业，同年进入广东省建筑设计研究院。是中国注册造价工程师、建筑工程概预算高级工程师。2009年起担任广东省建筑设计研究院造价中心副总造价师。2012年至2018年担任广州白云国际机场二号航站楼及配套设施造价专业负责人及审核人。负责各设计单位的造价专业及建设单位的概算总协调。

Graduated from the Department of Construction Management and Engineering of Chongqing Institute of Architecture and Engineering with a Bachelor's Degree in Engineering and joined GDAD in 1993. A registered cost engineer in China and a senior engineer in construction project budget. Deputy chief cost engineer in the Cost Center of GDAD since 2009. A budget principal and approver of T2 and its supporting facilities of the Baiyun Int'l Airport during 2012-2018.

许爱斌　Xu Aibin

许爱斌女士2007年福建工程学院工程管理工程系工学学士毕业，2008年进入广东省建筑设计研究院。是中国注册造价工程师、英国皇家特许测量师、建筑工程造价高级工程师。目前担任广东省建筑设计研究院造价中心副主任。2012年至2018年担任广州白云国际机场二号航站楼及配套设施造价专业负责人。

Graduated from the Engineering Management Department of Fujian Institute of Technology in 2007 with a Bachelor's Degree in Engineering; joined GDAD in 2008. A registered cost engineer in China, a Royal Chartered Surveyor, and a senior engineer in construction engineering. Deputy director of the Cost Center of GDAD. A budget principal of T2 and its supporting facilities of the Baiyun Int'l Airport during 2012-2018.

项目科研及成果

获奖记录

1. 2015年10月获国家三星级绿色建筑设计标识证书
2. 2015年6月获广东省优秀工程勘察设计奖BIM专项一等奖
3. 2017年11月获全国优秀工程勘察设计行业奖（市政给排水）一等奖
4. 2019年1月获SKYTRAX"全球五星航站楼"认证
5. 2019年3月获SKYTRAX"全球最杰出进步机场"及"中国最佳机场员工"奖
6. 2019年5月获第十三届第二批中国钢结构金奖
7. 2019年7月获广东省优秀工程勘察设计奖建筑工程一等奖
8. 2019年7月获广东省优秀工程勘察设计奖建筑装饰设计一等奖
9. 2019年7月获广东省优秀工程勘察设计奖建筑结构一等奖
10. 2019年7月获广东省优秀工程勘察设计奖建筑环境与设备一等奖
11. 2019年7月获广东省优秀工程勘察设计奖水系统工程专项一等奖
12. 2019年7月获广东省优秀工程勘察设计奖道路、桥梁、轨道交通一等奖
13. 2019年7月获广东省优秀工程勘察设计奖绿色建筑工程设计一等奖
14. 2019年7月获广东省优秀工程勘察设计奖建筑电气专项二等奖
15. 2019年7月获广东省优秀工程勘察设计奖园林和景观工程二等奖
16. 2019年8月获第十届中国威海国际建筑设计大奖银奖
17. 2019年11月获全国行业优秀勘察设计奖优秀（公共）建筑设计一等奖
18. 2019年11月获全国行业优秀勘察设计奖优秀绿色建筑一等奖
19. 2019年11月获全国行业优秀勘察设计奖优秀建筑结构二等奖
20. 2019年11月获全国行业优秀勘察设计奖优秀建筑环境与能源应用二等奖
21. 2019年11月获全国行业优秀勘察设计奖优秀水系统工程二等奖
22. 2019年11月获全国行业优秀勘察设计奖优秀建筑电气二等奖
23. 2019年11月获全国行业优秀勘察设计奖优秀市政公用工程设计三等奖

已获专利

1. "一种能够减轻自重且具有良好结构刚度的混合型楼盖"的发明专利
[专利号: ZL201521053110.0]
2. "灌注桩基施工期间同步进行的大直径灌注桩控壁岩体完整性探测方法"的发明专利
[专利号: ZL201610509714.4]
3. "一种用于种植大型乔木的梁柱节点"的实用新型专利[专利号: 201620423948.2]

试验及检测研究

1. 岩溶地区管波探测试验
2. 屋面抗风试验（包含抗风试验、疲劳试验、防水试验）
3. 万向支座试验
4. 膜结构铸钢节点试验、淋水试验
5. 钢管柱-混凝土土框架双梁节点设计试验
6. 大直径钢管混凝土柱混凝土浇筑及检测
7. 大直径钢管混凝土柱焊缝应力消除及检测
8. 建立了包含施工过程的长期健康监测体系，全面监控施工质量及后期健康监测，监测内容包含（1）应力应变（2）风压（3）温度（4）挠度（5）加速度与频率。
9. 广州白云国际机场屋面抗风揭试验报告
10. 广州白云国际机场钢管柱检测报告
11. 广州白云国际机场铸钢节点试验报告
12. 广州白云国际机场支座试验检测报告
13. 广州白云国际机场钢结构工程施工监测与健康监测报告

发表论文

建筑

1. 论文《大湾区机场航站楼设计的加速发展——以广州机场、深圳机场和珠海机场为例》，发表于2020年《世界建筑》第6期，陈雄
2. 论文《新岭南门户机场设计——广州白云国际机场二号航站楼及配套设施工程创作实践》，发表于2019年《建筑学报》第9期，陈雄、潘勇、周昶
3. 论文《广州白云国际机场二号航站楼公共空间照明设计》，表于2018年《云南建筑》第6期，陈雄、潘勇、钟世权、颜荣兴
4. 论文《大型航站楼建筑设计的多学科一体化设计——以新白云国际机场二号航站楼为例》，发表于2017年《城市建筑》第31期，陈雄、潘勇、赖文辉、郭其轶、易田、钟伟华
5. 论文《超大型航站楼设计实践与思考——广州新白云国际机场二号航站楼设计》，发表于2017年《建筑技艺》第11期，陈雄
6. 论文《大跨度建筑的形态与空间建构——以机场航站楼与体育馆为例》，发表于2016年《建筑技艺》第2期，陈雄
7. 论文《机场航站楼：非典型城市综合体设计研究》，发表于2016年《云南建筑》第3期，陈雄
8. 论文《高效·清晰·便捷——广州新白云机场T2航站楼旅客流程设计》，发表于2016年《建筑技艺》第7期，周昶
9. 论文《航站楼绿色建筑设计研究——以广州新白云国际机场二号航站楼为例》，发表于2014年《南方建筑》第3期，郭其轶
10. 论文《航站楼进出港旅客流程组织与登机模式设计研究——以广州新白云国际机场二号航站楼为例》，发表于2014年《a+a建筑知识》第6期，郭其轶

结构

11. 论文《Mechanical Behavior of Nine Tree- Pool Joints Between Large Trees and Buildings》，发表于2018年《KSCE Journal of Civil Engineering- eISSN》第22卷第8期，汪大洋、刘明琪、区彤、张永山
12. 论文《大直径钢管混凝土柱密实度检测及对接焊缝残余应力消减试验》，发表于2017年《建筑结构》第9期，张增球、区彤、谭坚
13. 论文《广州新白云国际机场二号航站楼混凝土结构设计》，发表于2016年《建筑结构》第21期，傅剑波、区彤、陈星、李桢章、戴鹏森、刘润富、刘振阳、卢德辉
14. 论文《广州新白云国际机场二号航站楼钢屋盖结构设计》，发表于2016年《建筑结构》第21期，张连飞、区彤、谭坚、陈星
15. 论文《广州白云国际机场二号航站楼预应力混凝土柱结构设计》发表于2016年《建筑结构》第S2期，谭和、李桢章、张鸿雁

机电

16. 论文《大型机场航站楼办票大厅照明设计》，发表于2019年《照明工程学报》第3期，钟世权
17. 论文《广州白云国际机场航站楼光伏发电项目设计》，发表于2019年《建筑电气》第4期，庄孙毅、钟世权、申雨佳
18. 论文《大型国际机场航站楼用电负荷研究》，发表于2019年《建筑电气》第2期，钟世权、何海平、蒋南雁、彭雪枫
19. 论文《交通建筑高大空间火灾探测器适用性分析》，发表于2018年《智能建筑电气技术》第6期，钟世权、周小蔚、申雨佳
20. 论文《浅谈航站楼强电系统设计》，发表于2018年《低碳世界》第7期，张文武
21. 论文《探析高大空间LED照明调光控制方案》，发表于2017年《江西建材》第9期，赵骥
22. 论文《浅谈一体化污水提升装置》，发表于2014年《中国给水排水》第8期，梁景晖

市政

23. 论文《广州白云国际机场扩建交通组织研究》，发表于2012年《建筑与文化》第12期，张正军

RESEARCHES & AWARDS

Awards

1. National Three-star Green Building Label Certification, Oct. 2015
2. The First Prize of 2015 Guangdong Engineering Exploration and Design Award (BIM), Jun. 2015
3. The First Prize of Guangdong Excellent Engineering Exploration and Design Award (Municipal Plumbing/Drainage), Nov. 2017
4. Rated as 5-star Terminal by SKYTRAX, Jan. 2019
5. World's Most Improved Airline and Best Regional Airport - China by SKYTRAX, Mar. 2019
6. The 13th Second Batch of Golden Prize Chinese Projects of Steel Structure, May 2019
7. The First Prize of Guangdong Excellent Engineering Exploration and Design Award (Building Engineering), Jul. 2019
8. The First Prize of Guangdong Excellent Engineering Exploration and Design Award (Building Decoration Design), Jul. 2019
9. The First Prize of Guangdong Excellent Engineering Exploration and Design Award (Building Structure), Jul. 2019
10. The First Prize of Guangdong Excellent Engineering Exploration and Design Award (Building Environment and Equipment), Jul. 2019
11. The First Prize of Guangdong Excellent Engineering Exploration and Design Award (Water System Engineering), Jul. 2019
12. The First Prize of Guangdong Excellent Engineering Exploration and Design Award (Road, Bridge, Rail Transit), Jul. 2019
13. The First Prize of Guangdong Excellent Engineering Exploration and Design Award (Green Building Engineering Design), Jul. 2019
14. The Second Prize of Guangdong Excellent Engineering Exploration and Design Award (Building Electrical Engineering), Jul. 2019
15. The Second Prize of Guangdong Excellent Engineering Exploration and Design Award (Garden and Landscape Engineering), Jul. 2019
16. Silver Prize of the 10th China Weihai International Architecture Design Award, Aug. 2019
17. The First Prize of National Excellent Engineering Exploration and Design Award (Excellent Architecture Design - Public Building), Nov. 2019
18. The First Prize of National Excellent Engineering Exploration and Design Award (Excellent Green Building), Nov. 2019
19. The Second Prize of National Excellent Engineering Exploration and Design Award (Excellent Building Structure), Nov. 2019
20. The Second Prize of National Excellent Engineering Exploration and Design Award (Excellent Build Environment and Energy Utilization), Nov. 2019
21. The Second Prize of National Excellent Engineering Exploration and Design Award (Excellent Plumbing System), Nov. 2019
22. The Second Prize of National Excellent Engineering Exploration and Design Award (Excellent Building Electrical System), Nov. 2019
23. The Third Prize of National Excellent Engineering Exploration and Design Award (Excellent Municipal Utilities), Nov. 2019

Patents

1. Invention patent for a hybrid floor covering capable of reducing dead weight and having good structural rigidity [Certificate No.: ZL201521053110.0]
2. Invention patent for detecting rock integrity of borehole wall of large-diameter cast-in-place pile along with the construction of cast-in-place pile foundation [Certificate No.: ZL201610509714.4]
3. Utility model patent for a beam-column joint for planting large arbors [Certificate No.: 201620423948.2]

Tests and Researches

1. Tube wave detection test in karst area
2. Roof wind resistance test (including wind resistance test, fatigue test, and waterproof test)
3. Universal bearing test
4. Membrane structure cast steel joint test, water spray test
5. Steel tubular column - concrete frame double beam joint design test
6. Concrete pouring and testing of large diameter concrete-filled steel tubular columns
7. Weld stress relief and inspection of large diameter concrete-filled steel tubular columns
8. The establishment of a long-term health monitoring system covering the construction process for comprehensive monitoring of construction quality and post-construction health monitoring, with contents monitored including (1) stress and strain (2) wind pressure (3) temperature (4) deflection (5) acceleration and frequency.
9. Guangzhou Baiyun International Airport roof wind resistance test report
10. Guangzhou Baiyun International Airport steel tubular column test report
11. Guangzhou Baiyun International Airport cast steel joints test report
12. Guangzhou Baiyun International Airport bearing test report
13. Construction monitoring and health monitoring report for Steel Structure Engineering of Guangzhou Baiyun International Airport

Publications

Architecture
1. Chen Xiong; Take Guangzhou Airport, Shenzhen Airport and Zhuhai Airport as Examples; *World Architecture*, 2020, (6)
2. Chen Xiong, Pan Yong, Zhou Chang; Designing the New Gateway Airport of South China: On the Practice of the Extension of No.2 Terminal and Supportive Facilities of Guangzhou Baiyun International Airport; *Architectural Journal*, 2019, (9)
3. Chen Xiong, Pan Yong, Zhong Shiquan, Yan Rongxing; Lighting Design for Public Space of Terminal 2 of Guangzhou New Baiyun International Airport; *Yunnan Architecture*, 2018, (6)
4. Chen Xiong, Pan Yong, Lai Wenhui, Guo Qiyi, Yi Tian, Zhong Weihua; Multidisciplinary Integrated Architectural Design of Large Terminal Buildings –A Case Study of Terminal 2 of New Baiyun International Airport; *Urban Architecture*, 2017, (31)
5. Chen Xiong; Design Practice and Thoughts on Mega Terminal Building - Design of Terminal 2 of Guangzhou New Baiyun International Airport; *Architecture Technique*, 2017, (11)
6. Chen Xiong; The Form and Space Construction of Long-span Building – A Case Study of Airport Terminal Building and Gymnasium; *Architecture Technique*, 2016, (2)
7. Chen Xiong; Airport Terminal: Design Study on Atypical Urban Complex; *Yunnan Architecture*, 2016, (3)
8. Zhou Chang; Efficient, Clear, Convenient - Passenger Flow Design for Terminal 2 of Guangzhou New Baiyun International Airport; *Architecture Technique*, 2016, (7)
9. Guo Qiyi; Study on Green Building Design of International Airport Terminal - A Case Study of Terminal 2 of Guangzhou New Baiyun International Airport; *South Architecture*, 2014, (3)
10. Guo Qiyi; Study on Flow and Boarding Design of Arrival and Departure Passengers in Terminals - A Case Study of Terminal 2 of Guangzhou New Baiyun International Airport; *a+a (Architecture & Art)*, 2014, (6)

Structure
11. Wang Dayang, Liu Mingqi, Ou Tong, Zhang Yongshan; Mechanical Behavior of Nine Tree-Pool Joints Between Large Trees and Buildings; *KSCE Journal of Civil Engineering-eISSN*, 2018,22(8)
12. Zhang Zengqiu, Ou Tong, Tan Jian; Experiment on Compactness Detection of Large Diameter CFST Column and Residual Stress Reduction of Butt Weld; *Building Structure*, 2017, (9)
13. Fu Jianbo, Ou Tong, Chen Xing, Li Zhenzhang, Dai Pengsen, Liu Runfu, Liu Zhenyang, Lu Dehui; Structural Design for Concrete System of Terminal 2 of Guangzhou New Baiyun International Airport; *Building Structure*, 2016, (21)
14. Zhang Lianfei, Ou Tong, Tan Jian, Chen Xing; Structural Design for Steel Roof of Terminal 2 of Guangzhou New Baiyun International Airport; *Building Structure*, 2016, (21)
15. Tan He, Li Zhenzhang, Zhang Hongyan; Structural Design for Pre-stressed Concrete Columns of Terminal 2 of Guangzhou Baiyun International Airport; *Building Structure*, 2016, S(2)

MEP
16. Zhong Shiquan; Lighting Design for Check-in Hall of Large Airport Terminals; *China Illuminating Engineering Journal*, 2019, (3)
17. Zhuang Sunyi, Zhong Shiquan, Shen Yujia; Design of PV Power Generation for Terminal Building of Guangzhou Baiyun International Airport; *Building Electricity*, 2019, (4)
18. Zhong Shiquan, He Haiping, Jiang Nanyan, Peng Xuefeng; Study on Electrical Loads of Large International Airport Terminal; *Building Electricity*, 2019, (2)
19. Zhong Shiquan, Zhou Xiaowei, Shen Yujia; Analysis on Applicability of Fire Detectors in Large and High Spaces of Transportation Buildings; *Electrical Technology of Intelligent Buildings*, 2018, (6)
20. Zhang Wenwu; Discussion on Design of HV System for Terminal Buildings; *Low Carbon World*, 2018, (7)
21. Zhao Ji; Study on Dimming Control of LED lights in Large Spaces; *Jiangxi Building Materials*, 2017, (9)
22. Liang Jinghui; Discussion on Integrated Sewage Lifting Equipment; *China Water & Wastewater Municipal Utilities*, 2014, (8)

Municipal Utilities
23. Zhang Zhengjun; Study on Traffic Organization for Expansion Project of Guangzhou Baiyun International Airport; *Architecture& Culture*, 2012, (12)

项目有关数据
FACTS & DATA

用地位置：	广州市北部，白云区人和镇与花都区新华镇的交界处	
占地面积：	1224919m² （机场总用地面积15km²）	
建筑面积：	航站楼65.87万m²，交通中心20.84万m²	
建筑密度：	22.58%	
建筑高度：	航站楼44.5m，交通中心12.95m	
地上层数：	航站楼4层，交通中心3层	
地下层数：	航站楼1层，交通中心2层	
设计容量：	年旅客吞吐量4500万人次，高峰小时客运量17299人次	
扩建机位：	78个（64个近机位，14个远机位）	
停车数量：	停车楼3769辆，室外1605辆	
绿地面积：	81164m²（航站楼屋顶绿化面积18526m²，道路及停车场绿化面积62638m²）	
外部装修：	航站楼：	墙面：低反射、低透视钢化玻璃、钢化中空/钢化中空彩釉玻璃、钢化中空夹胶/钢化中空夹胶彩釉玻璃、蜂窝铝板、铝单板
		屋面：铝镁锰合金直立锁边屋面板、局部钢筋混凝土绿化屋面、张拉膜（PTFE）
	交通中心：	墙面：金属网架、涂料
		屋面：钢筋混凝土绿化屋面、张拉膜（PTFE）
内部装修：	航站楼：	墙面：铝合金板、纤维增强硅酸盐板、瓷砖、不锈钢、木材、大理石、树脂板、涂料、钢化夹胶玻璃、工艺玻璃、墙布、墙纸、种植绿化墙面
		地面：花岗石、瓷砖、PVC防滑安全地板、地毯
		顶棚：铝合金板、微孔铝合金板、纤维增强硅酸盐板
	交通中心：	墙面：瓷砖、涂料
		地面：花岗石、瓷砖、固化地面
		顶棚：铝合金板、涂料
结构形式：	航站楼：	网架结构、预应力钢筋混凝土梁框架结构、预应力钢筋混凝土柱、钢管混凝土柱
	交通中心：	钢筋混凝土框架结构
基础形式及用量：	5124根钻（冲）孔灌注桩	
钢屋盖用钢量：	11960吨	
屋面压型钢板用钢量：	6197吨	
日供水量：	9256m³	
总用电量：	照明总电负荷24160kW，动力总电负荷66898kW	
空调总容量：	总冷负荷119553kW，总热负荷4080kW	
玻璃幕墙面积：	188800m²	
金属屋面板面积：	257200m²	
张拉膜面积：	26550m²	
办票岛总数：	12个	
柜台总数：	1288个	
行李提取盘总数及总长度：	总数21个，总长度2100m	
卫生间总数：	144个	
座椅总数：	17278位	
自动步道总数：	72部	
自动扶梯总数：	144部	
电梯总数：	118部	
设计周期：	9年4个月	
设计日期：	2006年8月~2015年12月	
施工周期：	4年6个月	
施工日期：	2013年8月~2018年2月	
竣工日期：	2018年2月6日	
运营日期：	2018年4月26日	

Location: At the junction between Renhe Town of Baiyun District and Xinhua Town of Huadu District in the north of Guangzhou
Site area: 1,224,919m² (airport site: 15km²)
Building size: T2: 658,700m², GTC: 208,400m²
Building coverage: 22.58%
Building height: T2: 44.5m, GTC: 12.95m
Aboveground floors: T2: 4, GTC: 3
Underground floors: T2: 1, GTC: 2
Design capacity: Annual passenger traffic: 45 million, peak-hour passenger traffic: 17,299
New stands: 78 (64 contact, 14 remote)
Parking spaces: Parking building: 3,769, outdoor: 1,605
Greening area: 81,164m² (T2 roof: 18,526m², roads and parking lots: 62,638m²)
Exterior finish: T2: Wall: low-reflection and -transmission tempered glass, tempered insulated glass/tempered insulated fritted glass, tempered insulated laminated/tempered insulated laminated fritted glass, honeycomb aluminum plate, aluminum panel
Roof: Al-Mg-Mn alloy vertical edged roof board, partially reinforced concrete greening roof, tensile membrane (PTFE)
GTC: Wall: metal grid, paint
Roof: reinforced concrete greening roof, tensile membrane (PTFE)
Interior finish: T2: Wall: aluminum alloy plate, fiber reinforced silicate plate, ceramic tile, stainless steel, timber, marble, resin board, paint, tempered laminated glass, art glass, wall fabric, wall paper, greening wall
Floor: granite, ceramic tile, PVC skid-resistant floor, carpet
Ceiling: aluminum alloy plate, micro-perforated aluminum alloy plate, fiber reinforced silicate plate
GTC: Wall: ceramic tile, paint
Floor: granite, ceramic tile, curing floor
Ceiling: aluminum alloy plate, paint
Structure: T2: grid structure, prestressed reinforced concrete beam frame structure, prestressed reinforced concrete column, concrete filled steel tubular column
GTC: reinforced concrete frame structure
Foundation: 5,124 bored piles
Steel roof: 11,960 tons of steel; roof profiled sheet: 6,197 tons of steel
Daily water supply: 9,256m³
Power consumption: total lighting load: 24,160kW, total power load: 66,898kW
AC capacity: total cooling load: 119,553kW, total heating load: 4,080kW
Glass curtain wall area: 188,800m²
Metal roof board area: 257,200m²
Tensile membrane area: 26,550m²
Check-in islands: 12
Counters: 1288
Baggage carousels: Qty 21; total length 2100m
Washrooms: 144
Seats: 17,278
Automatic people movers: 72
Escalators: 144
Elevators: 118

Design: 9 years + 4 months
Period: Aug. 2006 - Dec. 2015
Construction: 4 years + 6 months
Period: Aug. 2013 - Feb. 2018
Completion: Feb. 6, 2018
Operation: Apr. 26, 2018

本书编辑团队
EDITORIAL TEAM

主编：	陈雄、潘勇、周昶
执行编辑：	郭其轶、金少雄
文字编辑：	赖文辉、易田、钟伟华、杨坤、区彤、傅剑波、李恺平、谭和、劳智源、钟世权、赵骥、陈小辉、郭林文、梁景晖、赖振贵、李东海、陈伟、刘明、黎洁、胡智敏、饶欣频、姜颐雯
图纸编辑：	许尧强、易田、倪俍、戴志辉、魏画野、张栋、黄河清、郑重第、许秋滢、马智超
摄影：	潘勇（负责本书除以下注明作者外的其他所有照片的拍摄）； 傅兴P76、P124右下角、P147、P153、P155上方、P158右下角、P182左下角、P194-195、P209； 李开建P145、P151、P154左下角、P170下方、P182右下角2张、P183
翻译：	梁玲、张丽娟
图书设计：	廖荣辉
封面设计：	潘勇

Chief Editors: Chen Xiong, Pan Yong, Zhou Chang
Executive Editors: Guo Qiyi, Jin Shaoxiong
Text Editors: Lai Wenhui, Yi Tian, Zhong Weihua, Yang Kun, Ou Tong, Fu Jianbo, Li Kaiping, Tan He, Lao Zhiyuan, Zhong Shiquan, Zhao Ji, Chen Xiaohui, Guo Linwen, Liang Jinghui, Lai Zhengui, Li Donghai, Chen Wei, Liu Ming, Li Jie, Hu Zhimin, Rao Xinpin, Jiang Yiwen
Drawing Editors: Xu Yaoqiang, Yi Tian, Ni Liang, Dai Zhihui, Wei Huaye, Zhang Dong, Huang Heqing, Zheng Zhongdi, Xu Qiuying, Ma Zhichao

Photographers: Pan Yong (all the photos except those indicated with other names): Fu Xing: P76, the lower right corner of P124, P147, P153, the upper part of P155, the lower right corner of P158, the lower left corner of P182, P194, P195, P209; Li Kaijian: P145, P151, the lower left corner of P154, the lower part of P170, two pieces in the lower right corner of P182, P183

Translators: Liang Ling, Zhang Lijuan
Graphic Designer: Liao Ronghui
Cove Designer: Pan Yong

编后记

2018年4月26日，白云机场二号航站楼启用，这是一个非常值得纪念的日子！令人欣喜的是，自通航以来，二号航站楼获得了社会各界、民航领域与建筑设计业界的广泛好评。设计工作2006年开始，至启用刚好十二年一个轮回，而从1998年开始为新白云机场服务算起，则至启用刚好整整二十年！此时此刻，需要充分展现二号航站楼的风采，需要全方位进行设计总结。于是，我们决定为二号航站楼出一本专著，正如2006年为一号航站楼设计出版专著一样。

我们制定了新的编辑方针，专业齐全系统完整而又突出重点，图文并茂地展现设计成果，充分阐述设计理念、技术创新和设计特色。

在这里我们要特别感谢华南理工大学建筑设计研究院董事长何镜堂院士在百忙之中为本书作序！当年，也是承蒙何院士为一号航站楼专著作序。

感谢项目组的各专业多位同事编写文稿、整理图纸和校对文稿，特别要感谢郭其轶、金少雄两位同事，他们为本书的编辑出版做了大量具体细致的工作，从全书结构拟定、资料选择到试排版，反复调整优化，务求保证品质。

廖荣辉先生在十几年前为一号航站楼专著排版，现在又为本书排版和装帧设计。他不辞劳苦，反复修改调整，依然保持追求完美的专业精神。感谢潘勇总建筑师、傅兴先生和李开建先生拍摄的作品的照片。感谢梁玲女士团队为本书所做的全部翻译工作。

时光飞逝，回想起1998年元旦伊始，白云机场业主为新白云机场举行国际竞赛，多家全球著名机场设计公司参加，GDAD独立参加了这次国际竞赛。在1998年的金秋十月，GDAD被白云机场业主确定为新机场的中方设计单位，从此，一个光荣的使命等待广东省建筑设计研究院去完成！

二十载时光，GDAD陪伴新白云机场经历了三个发展阶段，包括1998~2004年的机场迁建，2005~2009年的一号航站楼扩建，2006~2018年的二号航站楼新建。

二号航站楼位于一号航站楼北侧，设计年旅客量为4500万人次，总机位共78个。第一期工程建设主楼及东、西各两条指廊及相关配套工程，总建设规模接近90万m²，是目前我国运行的规模最大、功能最齐全的航站楼综合体之一。

二号航站楼拥有流畅的旅客流程和完善的功能设施；建构与一号航站楼和谐一致的建筑造型；保留弧线形的主楼和人字形柱及张拉膜雨篷这些特有元素；体现岭南地域特色的花园空间及装修设计；强化商业资源整合为提高非航收入奠定基础；增加设计弹性应对未来需求变化；注重绿色环保节能设计，达到中国绿色建筑三星标准；成为展示公共艺术的枢纽门户；反映当今中国最新的建筑技术水平。

在过去的十二年间，设计经历了几次重大修改。GDAD始终高度重视，投入了大量的人力和心血，设计团队克服了各种各样数不清的困难，解决了非常多的技术难题，为了完成这项宏大的工程，一直努力拼搏，兢兢业业，希望尽可能减少遗憾，为广东省广州市奉献一个全新的先进的门户航站楼，实现了从合作设计到主创设计的飞跃！

二号航站楼是迄今GDAD设计的单项规模最大、综合难度最高、投入人力最多的超大型公共建筑项目，代表了GDAD最新最高的建筑水平，全院高度重视并给予了大力支持。在这里，特别感谢历届院领导班子的信任和支持！特别感谢机场设计研究所/ADG建筑创作工作室牵头进行项目的原创设计，对项目做出了突出贡献！感谢GDAD其他参与部门的辛勤付出和积极贡献，包括第二建筑设计研究所、第三机电设计研究所、第一市政工程设计研究所、第二机电设计研究所、第一机电设计研究所、第一城市与景观设计研究所、第二市政工程设计研究所、造价中心以及绿色建筑设计研究中心。感谢相关合作单位的大力支持与协作配合。

广州白云国际机场是南方航空总部基地，承担起祖国南大门连接欧、美、澳、非的空中桥梁。崭新的二号航站楼与一号航站楼交相辉映，助力广州建设粤港澳大湾区核心城市以及中国建设成为世界民航强国的伟大战略目标！

真情二十载，助白云腾飞！借此本书出版之际，我们再次衷心感谢白云机场业主、南航、联检单位、各参建单位给予的信任、理解与大力支持！我们期待未来继续共同努力，为白云机场的建设谱写新的绚丽的篇章！

AFTERWORD

April 26, 2018 was an extremely memorable day as it saw the grand opening of the Terminal 2 (T2) in the Baiyun Airport. Much to our delight, the newly opened T2 was widely acclaimed by the general public as well as professionals from the civil aviation and architectural design industry. The opening of T2 coincided with the 12th anniversary of the terminal design since 2006, and the 20th anniversary of our cooperation with the new Baiyun Airport since 1998. In this context, we deem it the right timing to present this fascinating project with a comprehensive recap of the project design. Therefore, we decided to publish a book for T2, just as we did for Terminal 1 (T1) in 2006.

With the newly established guidelines, this book intends to offer a full picture of the terminal design, covering all disciplines and systems in a comprehensive and focused manner while vividly presenting the design results, concepts, highlights and technological innovation via both narratives and images.

Here we would like to express our heartfelt thanks to Academician He Jingtang, president of the Architectural Design & Research Institute of SCUT, for sparing time from his busy schedule to write a foreword for this book, just as he did years ago for the book of T1.

Our thanks also go to the multidisciplinary project team for preparing and reviewing the drawings and texts in this book. In particular, we wish to thank Guo Qiyi and Jin Shaoxiong for the strenuous efforts they have exerted on this book. From the overall framework of the book, to its specific contents and trial typesetting, they've made rounds of revisions and refinements to ensure the high quality of the book.

More than a decade ago, Mr. Liao Ronghui provided professional graphic design for the book of T1. Now with the same professionalism and meticulous work attitude, he completes the graphic design for this book. Last but not least, we wish to deliver our thanks to chief architect Pan Yong, Mr. Fu Xing and Mr. Li Kaijian for the project photos in this book, and Ms. Liang Ling and her team for the English translation of this book.

How time flies and I can still recall that, right after the New Year's Day of 1998, the client launched an international design competition for the new Baiyun Airport, attracting quite a few world-renowned airport design firms and GDAD, an independent participant. In October 1998, the client announced GDAD the local design firm for the new airport. From then on, a grand mission was established for GDAD to accomplish.

Over the past 20 years, GDAD had witnessed three phases of airport development, namely the airport relocation from 1998 to 2004, T1 expansion from 2005 to 2009, and T2 development from 2006 to 2018.

Located to the north of T1, T2 is planned with an annual passenger handling capacity of 45 million people and a total of 78 stands. Phase I of the Project, with a GFA of nearly 900,000m² for the main building, two piers in the east and west, and related supporting facilities, boasts one of the largest terminal complexes with the most sophisticated functions in operation in China.

T2 features smooth passenger flow and fully-fledged functional facilities. It maintains a harmonious and consistent architectural style with T1, and retains the unique airport elements such as the curvy main building, the herringbone columns and tensile membrane canopy. Other design highlights include: garden spaces and interior design of typical Lingnan characteristics, integration of commercial resources to generate more non-airline revenues, design flexibility to accommodate future demands and changes, green, environmental-friendly and energy-efficient design to be labeled as a 3-star Green Building in China, and the vision to make T2 a gateway hub of the public art and a showcases of the country's latest building technologies.

Over the past 12 years, the project has seen several rounds of major design changes which were addressed by GDAD with great attention and numerous efforts. Overcoming great difficulties and tackling various technical challenges, the design team has been working diligently on the project to present the city with a brand new gateway terminal and eventually evolve from a cooperative designer to chief designer!

T2 is the largest, the most sophisticated and the best-staffed single mega-sized public building GDAD has ever worked on. It represents the highest expertise of GDAD and received great supports from all GDAD departments. Our special thanks go to all GDAD leaderships for their trust in and support to us, and Airport Design Group (ADG)/ADG Architectural Design Studio for their leading role in creative architecture design and outstanding contribution to the project. Our thanks also go to other GDAD departments for their hard work and meaningful contribution, including the 2nd Architecture Studio, the 3rd E&M Studio, the 1st Municipal Utility Studio, the 2nd E&M Studio, the 1st E&M Studio, the 1st Urban Design and Landscape Studio, the 2nd Municipal Utility Studio, the Cost Engineering Center and the Green Building Design and Research Center. Last but not least, we are also very thankful to all our project partners for their great support and cooperation.

Guangzhou Baiyun International Airport, as the headquarters base of the China Southern Airlines (CSN), serves as an air bridge to connect Guangzhou, known as China's southern gateway, with Europe, the United States, Australia and Africa. The newly built T2 well complements with T1, and will make its due contribution to the grand strategic goal of making Guangzhou a core city in GBA and China a leader in world's civil aviation industry.

The past two decades is a witness to our efforts devoted to the success of the airport. On the special occasion of the publication of this book, we would like to once again express our sincere thanks to the client, CNS, CIQ authorities, and all stakeholders for the trust, understanding and great support rendered to us! We look forward to working with you toward a more promising future of the Baiyun Airport.

图书在版编目(CIP)数据

广州白云国际机场二号航站楼及配套设施 / 陈雄，潘勇，周昶主编. 北京：中国建筑工业出版社，2019.8
ISBN 978-7-112-24057-9

Ⅰ.①广… Ⅱ.①陈… ②潘… ③周… Ⅲ.①国际机场－航站楼－配套设施－建筑设计－广州 Ⅳ.① TU248.6

中国版本图书馆 CIP 数据核字 (2019) 第 161309 号

责任编辑：唐旭　吴绫
文字编辑：李东禧　孙硕
责任校对：李欣慰

广州白云国际机场二号航站楼及配套设施
主编：陈雄、潘勇、周昶
*
中国建筑工业出版社出版、发行（北京海淀三里河路 9 号）
各地新华书店、建筑书店经销
恒美印务（广州）有限公司印刷
*
开本：787×1092 毫米　1/12　印张：22　字数：832 千字
2020 年 6 月第一版　2020 年 6 月第一次印刷
定价：258.00 元
ISBN 978-7-112-24057-9
（34560）

版权所有　翻印必究
如有印装质量问题，可寄本社退换
（邮政编码 100037）